Politics and Violence in Cuban and Argentine Theater

Katherine Ford

palgrave
macmillan

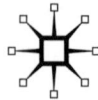

POLITICS AND VIOLENCE IN CUBAN AND ARGENTINE THEATER
Copyright © Katherine Ford, 2010.

All rights reserved.

First published in 2010 by
PALGRAVE MACMILLAN®
in the United States—a division of St. Martin's Press LLC,
175 Fifth Avenue, New York, NY 10010.

Where this book is distributed in the UK, Europe and the rest of the world, this is by Palgrave Macmillan, a division of Macmillan Publishers Limited, registered in England, company number 785998, of Houndmills, Basingstoke, Hampshire RG21 6XS.

Palgrave Macmillan is the global academic imprint of the above companies and has companies and representatives throughout the world.

Palgrave® and Macmillan® are registered trademarks in the United States, the United Kingdom, Europe and other countries.

ISBN: 978–0–230–61314–0

Library of Congress Cataloging-in-Publication Data

Ford, Katherine.
    Politics and violence in Cuban and Argentine theater / Katherine Ford.
      p. cm.
    Includes bibliographical references and index.
    ISBN 978–0–230–61314–0
      1. Cuban drama—20th century—History and criticism. 2. Argentine drama—20th century—History and criticism. 3. Violence in literature. 4. Politics and literature—Cuba—History—20th century. 5. Politics and literature—Argentina—History—20th century. I. Title.

PQ7381.F67 2010
862'.6093581—dc22
                                                         2009023753

A catalogue record of the book is available from the British Library.

Design by Newgen Imaging Systems (P) Ltd., Chennai, India.

First edition: February 2010

10 9 8 7 6 5 4 3 2 1

Printed in the United States of America.

POLITICS AND VIOLENCE IN CUBAN AND
ARGENTINE THEATER

For Rufino

Contents

Preface: Understanding the Place of Theater in Spanish America   ix

Acknowledgments   xxi

Introduction: Difficult Times: Considering Dramatic Violence   1

1. Who's Afraid of Virgilio Piñera? Violence and Fear in *Dos viejos pánicos* (1968)   25

2. Cobwebs of Memory: History Made with Violence in Abelardo Estorino's *La dolorosa historia del amor secreto de don José Jacinto Milanés* (1974)   57

3. Filming the Bourgeoisie: Defining Identity with Violence in Eduardo Pavlovsky's *La mueca* (1970)   97

4. Disorderly Conduct: The Violence of Spectatorship in Griselda Gambaro's *Información para extranjeros* (1973)   137

Conclusion: Transforming Spectacles   175

Notes   181

Bibliography   205

Index   215

Preface: Understanding the Place of Theater in Spanish America

Spectacle and performance are defining elements of human existence and no geographical or historical context is free of theater and its characteristics. Spanish America enjoys a rich tradition of theater written and produced from the days of the colony up until the present. While this theater has often been in dialogue with European traditions, it is a mistake to dismiss it as simply a copy of theater from other regions. By introducing here some of the tendencies that have defined Spanish American theater up to the middle of the twentieth century, this preface introduces the reader to what was happening onstage and around the theater of this region.

While not much is known of the theatrical texts that may have existed before the arrival of the Spanish to the Americas, we can speak of the role of spectacle and performance in some indigenous cultures. Spectacle formed a central role in these early encounters and helped the Spanish gain control over the indigenous people that they subjugated. Adam Versényi in *Theatre in Latin America: Religion, Politics, and Culture* discusses the importance that spectacle played in Hernán Cortés' actions as he arrived to the New World, both with the indigenous groups he conquered and the Franciscan friars sent by the Spanish crown to evangelize, underlining the role that spectacle played in gaining control in the Spanish colonies.[1] For Cortés, conquering New Spain was not simply a military feat but a political one as well that would be won by manipulating images and performances. Indeed, as Diana Taylor outlines in *Theatre of Crisis*, for indigenous cultures such as the Aztec, the Maya, and the Inca, "Spectacle was power."[2] This meant that the creation of spectacle—both political and theatrical—took center stage in the unfolding drama of the creation of a colony. In the religious realm, intertwined with the political in the early days of the colony, the Church took advantage of the existing use of spectacle and theatricality of the Aztec world to Christianize the indigenous peoples. Dating to before the arrival of the Europeans, Aztec warfare was composed of ritualized activities whose object was to capture prisoners who would be sacrificed in another regulated ritual. These ritualized activities allowed the Church, another proponent of ritual, to come in and evangelize through the use of religious representations in which the vanquished indigenous would participate. Spectacle and theater, then, defined the past and laid out the future of the people and the land of the Americas.

Later on during the days of the colony, the majority of theater was often that of the Spanish baroque masters, such as Lope de Vega, Calderón de la Barca, and Tirso de Molina. This was not the only theater since Spanish America did produce its own playwrights who often blended European traditions with elements of the indigenous cultures that surrounded them. Sor Juana Inés de la Cruz (1651–1695) is, without a doubt, the foremost Spanish American playwright of the colonial time, though she is equally well-known for her poetry and her essays. Born in Mexico, she entered the convent early in her writing career in order to ensure her right to study. The breadth of her theatrical work is large and her work is among the best. Sor Juana's theater is both within the Spanish baroque tradition and a challenge from the periphery to the dominance of the center, because she is a woman writing when men controlled cultural production and for the fact that she wrote within the Spanish tradition from a convent in Mexico City.

Sor Juana, however, is not the only colonial to penetrate the center of Baroque theater. Juan Ruiz de Alarcón (1581–1639) is considered one of the major dramatists of the Golden Age in Spain, the country to which he emigrated from Mexico as a young man. Though he wrote in and set his plays in Spain rather than the New World and is often for that reason included in the names of peninsular playwrights, he was born in the New World. He is most known for *La verdad sospechosa*, published in 1634. Ruiz de Alarcón's inclusion in the canon of the Spanish Golden Age is often not questioned. However, some critics also include him within the list of Colonial writers, an issue that Alberto Sandoval Sánchez argues in his essay on Ruiz de Alarcón.

Whereas seventeenth-century theater mostly continued a peninsular tradition (though with New World innovation) the eighteenth century, marked by neoclassicism in the literary realm, began to see independence movements throughout Latin America, most especially towards the end of the century. In theater, this can be seen in the increased use of indigenous elements on the Spanish American stage. There is one particularly important example of this from Peru. The play *Ollantay* was originally written in Quechua and early manuscripts date from the eighteenth century. It was performed in 1780 for the indigenous leader Condorcanqui and recounts a love story between a warrior and an Incan princess, the warrior's punishment and his consequent rebellion. After his defeat ten years later, the warrior and the princess are pardoned and reunited with their daughter.

There are links between this play and an indigenous rebellion against the Crown that suggest growing social tensions. In the realm of theater, the play integrates elements from the Spanish and the indigenous worlds.

In a more officially sanctioned move, in the second half of the eighteenth century, numerous large theaters were built in the important cities throughout Spanish America, such as Buenos Aires' Teatro de la Ranchería or Casa de Comedias (which opened in 1789 but was destroyed by fire in 1792). This gave both traveling theater groups and local playwrights a new professional space that was dedicated to theater production, though this was not necessarily available to all, both in terms of financial access and political connections. Nevertheless, these new spaces legitimated these destinations and the theater within them. This can be seen as a move towards independence and national definition that would mark the nineteenth century.

Theater of the nineteenth century continued the tendency that connected European traditions with elements that were distinctly American. Theater and performance were extremely popular and not limited to the traditional idea of the stage. Instead, we also see spectacle centered around spaces such as the circus and more popular places, such as the *chingana* in Chile, an inn, restaurant or café with singing and dancing that was often a site that promoted new republican sentiments.[3] In addition, the theater of many of the countries of Spanish America began to distinguish itself and take on specific national characteristics. In Cuba, this can be seen in the *teatro bufo*, a definitively Cuban genre of theater which emerged in the late 1860s. It used the figure of the *negrito* and was influenced in part by the traveling minstrel shows from the United States. The *negrito* was a white performer in black face who became a beloved figure on the stage and came to represent Cuba and Cubanness. Jill Lane's *Blackface Cuba, 1840–1895* details how the *negrito*, blackface, and *teatro bufo* contributed to the evolution of the Cuban stage in the nineteenth century and beyond. She asserts that this humor functioned at two levels: that of controlling blackness and of negotiating whiteness within the colonial hierarchy.[4]

In the middle of the nineteenth century in Cuba, José Jacinto Milanés (1814–1863) (whose work will be discussed in chapter two) premiered his *El conde Alarcos* (1838) in the Teatro Tacón in Havana, considered by many to be the beginning of the Cuban stage. This play, as we will see, like much of the theater written at the time, had strong nationalistic characteristics. Along these same patriotic calls for independence, José Martí also wrote and published the one-act play *Abdala* in 1869, with a black soldier in the role of hero. Both the themes and the hero of Martí's play make it distinctly Cuban and show the innovation that marked the theater and how theater aimed to change the societies from which it came.

In the Río de la Plata area, during this same period, immigration and the tension between the metropolis and the country were some of the topics that defined the area and its art. In the second half of the nineteenth century the representations of the gaucho and his troubles dominated the stages. *Juan Moreira*, originally a novel (1884/1886) by Eduardo Gutiérrez (1853–1890), was adapted for the circus. Its immense popularity led to its introduction onto the stage, arriving in Montevideo in 1889 and Buenos Aires in 1891. As Versényi points out, this marked the beginning of the end of the *teatro gaucho* but signaled an important contribution to Spanish American theater.[5]

The Uruguayan Florencio Sánchez (1875–1910), on the other hand, turned his attention to the complexities of the changing world and the shift to modernity in plays such as *Barranca abajo* (1905), another common topic in the theater of the *rioplatense* area. In *Barranca abajo*, Sánchez highlighted the ending of a lifestyle and a shift from the rural to an increased focus on an urban world. The play is a Latin American tragedy where don Zoilo loses his land through the courts and his family to death and changing morals, all leading to his suicide.

Innovations continued to occupy an important space in the theatrical production of the 1920s and 30s in Spanish America, and highlighted the role of political thought and artistic innovation through theater.[6] Popular theater groups and playwrights played a pivotal role in bridging traditional paradigms with new forms of theatrical production and political debate, like their predecessors. The tension between the old and the new is defined even more sharply in the work of Armando Discépolo of Argentina (1887–1971) and the *grotesco criollo*. His play *Stefano* (1928) is a particularly pertinent example of the innovations of theater found in Argentina in the 1920 and 30s. The *grotesco criollo* focused on the new immigrants that were forming a part of Buenos Aires and on the social hardships and economic poverty that they suffered. It is closely identified with the lower classes, often comprised of immigrant communities. *Stefano* portrays a family of Italian immigrants in Buenos Aires who come looking for a better life but are unable to find it due to economic and social conditions. Discépolo's work focused on the contribution of immigration to national identity and the ways that the city and the nation fail these people, factors that would define Argentina in the years to come.

These questions and crises of identity and the self were not limited to South America. *El gesticulador* (1938) from Rodolfo Usigli of Mexico (1905–1979) questions the roles of men within the new ideal created by the Mexican Revolution. This play about a supposed imposter of a revolutionary hero highlights the struggles of the new society and reveals the fissures of where reality falls below the ideals. This canonical example shows how theater entered into the public debates about this new Mexico and what it meant to be Mexican and how far the country had deviated from its revolutionary goals.

Xavier Villaurrutia (1903–1950) of Mexico is another central contributor to a theater that was exerting its own new, national definitions. Villaurrutia is an important figure in the formation of theater groups, having founded the group *Ulises* (with Celestino Gorostiza (1904–1967)) in 1928 and then *Teatro Orientación* in 1932. Both of these groups wanted to transform the idea of theater by breaking its connection to earlier, outside models. In both his work with these groups and in his own dramatic writing, Villaurrutia contributed to an innovation in the idea and definition of Mexican theater. Both he and Usigli would have a profound effect on the work of the future playwrights of their country, such as can be seen in the work of Emilio Carballido (1925–2008), whose extensive work defies simple classification.

While these examples of theater considered topics specifically focused on their respective countries, their and others' innovation on and around the stage continued as the century progressed and the theater of these years can be characterized as diverse and wide-reaching. This diversity can be seen in the influence of the Theater of the Absurd, a movement that originated in Paris in the 1940s and 1950s. The Theater of the Absurd was born from a desire to reform theater in the wake of the destruction of World War II and its aftermath. The notion of the absurd as a way to understand the post-war situation of humankind originates out of Albert Camus (1913–1960) and Jean Paul Sartre (1905–1980), who believed that Man should accept the ultimate lack of meaning to the world. Antonin Artaud (1896–1948) and Martin Esslin (1918–2002) shed light on what they saw as the absurdity of the world. Artaud, whose Theater of Cruelty exchanged many ideas with the Theater of the Absurd, and Esslin share the belief that theater should provoke the spectator beyond his/her expectations and to innovate what happens both onstage and offstage.[7] Theater of Cruelty is a concept from Artaud that describes a type of theater where traditional authority vanishes. The cruelty to which the name refers, "is not the cruelty we can exercise upon each other by hacking at each other's bodies […] but the much more terrible and necessary cruelty which things can exercise against us."[8]

Theater, for Artaud and the absurdists, was not to be an escapist experience that took one away from daily concerns, but had as its mission to awaken the spectator to the world and its realities. This disillusionment with theater was manifested in opposition to the classics such as Shakespearean and Romantic theater and attempted to uncover the ultimate absurd nature of a world rendered inexplicable by contemporary life.

Eugene Ionesco's *The Bald Soprano* (1950) and Samuel Beckett's *Waiting for Godot* (1953) are classic examples of how the Absurd portrayed the anxiety towards life and death and the ultimate irrationality that these writers believed characterized the world.

The ideas that the playwrights and theorists of the Theater of the Absurd in Europe put forth were quickly adapted by many dramatists in Latin America, as will be explored in Chapter one in reference to Virgilio Piñera. George Woodyard, in his 1969 essay "The Theatre of the Absurd in Spanish America," explores the employment of ideas from the European Theater of the Absurd in the context of 1960s Spanish America. For Woodyard, the use of the Absurd and the number of absurd plays in the Americas differs given the change of context, though the focus on irrationality and fragmentation remains central. He identifies the most important elements of the Spanish American Theater of the Absurd as the following: plays with two characters, anti-heroes, physical violence stemming from feelings of contempt and hatred, and an insistence on fragmentation.[9]

While there is a rich history of the Absurd influencing Spanish American theater, it is unquestionable that the theater artists in Latin America have innovated the Absurd to make it their own and tailor it to their unique circumstances. Woodyard details many of the important voices that were writing at the time who were influenced by the Absurd. These names are also ones that would remain in the forefront of theater in subsequent years and some that will be studied in more detail later in this book, such as Virgilio Piñera of Cuba and Griselda Gambaro of Argentina. In addition to these two, Woodyard highlights the particularly Absurd identifications in the theater of Elena Garro (1920–1998) of Mexico, Cuba's Antón Arrufat (b. 1935), and Jorge Díaz (1930–2007) of Chile. Díaz is considered, without a doubt, one of the central names of the Absurd in Spanish America, particularly with his *El cepillo de dientes*, first produced in 1961. In this play we see various characteristics of the Absurd, such as a cyclical structure, irrational dialogue, generic characters, and a gratuitous violence.

While Woodyard uses the term Absurd to talk about this theater, some critics have preferred to call this theater Absurdist in order to mark the differences between the European examples and those of Spanish America. The main difference between these two that has been identified by such critics as Daniel Zalacaín and Raquel Aguilú de Murphy is in the political-social topics, though the same alienation is explored on both sides of the Atlantic.[10] Woodyard and others, such as Terry Palls and Eleanor Jean Martin, have not seen the need to differentiate between the two manifestations, though they do recognize the strong social-political aspect of the Spanish American plays.[11]

While this difference of terminology is not, in my opinion, a strong point of contention in interpretation, it does underline the importance of the Absurd and its influences in Spanish American theater, an importance that goes beyond those playwrights that are strictly identified as Absurd. Woodyard identifies five other playwrights that, though perhaps not strictly Absurd, were unquestionably influenced by the avant-garde movements of the time, the Absurd one of them. These playwrights include the names of Osvaldo Dragún (1929–1999) of Argentina, Puerto Rico's René Marqués (1919–1979), Carlos Solórzano (b. 1922) of Guatemala, Agustín Cuzzani (1924–1987) of Argentina, and Mexico's Emilio Carballido. The Absurd can be seen to permeate the work of countless playwrights, including many of those studied here and will be examined further in the chapter on Virgilio Piñera.

Bertolt Brecht's innovative theories on theater similarly helped to revitalize theater and influence dramatic production in Spanish America, as Fernando de Toro details in his *Brecht en el teatro hispanoamericano*.[12] Perhaps the two most important concepts from Brecht that informed Latin American playwrights at this time are *Verfremdung* (known as alienation or distancing) and *episches Theater* (epic theater). Brecht's goals were to awaken the spectators to the situation before them and to motivate them to rationally consider what was happening. He saw theater as the means to inspire the spectator to new thoughts that would in turn renovate theater itself as well as the outside world: "We need a type of theatre which not only releases the feelings, insights and impulses possible within the particular historical field of human relations in which the action takes place, but employs and encourages those thoughts and feelings which help transform the field itself."[13] This transformation of the field and the community in which it interacts was brought across through the use of alienation. Brecht considered that when the spectators are discouraged from identifying with the main characters they would be able to reflect more freely on the material before them: "A representation that alienates is one which allows us to recognize its subject, but at the same time makes it seem unfamiliar."[14]

Both influenced by and shaping German and international theater in the 1920s and beyond, Brecht's epic theater is concerned, in the words of John Fuegi, "with rawness, with facts, with bringing the whole world into theater in order to give the public lessons on political and economic questions of the day."[15] Three of the central initiatives of epic theater were the use of new and different materials, a production style that underscored reason over emotion, and the creation of a new type of spectator who would be able to coolly appreciate this theater.[16] Brecht advocates with these ideas a renovation of theater that calls for a stronger and more engaged spectator, one who will carefully consider the represented material in the light of historical and political events.

In the 1950s and especially the 1960s, the increased production of these types of experimental theater highlighted the connection between popular theater groups and the political and social context in which they operated. Nora Eidelberg explores the manifestations of experimental theater in her *Teatro experimental hispanoamericano, 1960–1980: La realidad social como manipulación*.[17] Along this same time period, Nuevo Teatro Popular (New Popular Theatre) was starting to consolidate its definition into one that could be connected across various countries of Latin America. Its peak can be seen around 1965–1975, according to *Latin American Popular Theatre: The First Five Centuries* from Judith A Weiss, et al. While theater that can in retrospect be defined as popular has existed throughout the modern history of Spanish America, the emergence of what is referred to as Nuevo Teatro Popular materialized in connection with other social and political developments of the 1960s and 1970s, a time of tumultuous change and innovation that contributed to this New Popular Theatre. Both grass-roots initiatives and more professionally trained, New Popular Theatre tended to be organized around groups or collectives more than exclusively individuals that came together and apart based on the event. Nevertheless, these groups were often closely identified with a founder or director, such as Enrique Buenaventura (1925–2003) and the Teatro Experimental de Cali (TEC) of Colombia, perhaps the most well-known experimental theater group from Spanish America. The topics tended to be wide-ranging and tied to class and cultural identity, themes that would empower their audiences. These groups connected with one another through festivals dedicated to popular theater and helped to create this genre into a movement that spanned national and cultural borders. Buenaventura, best known for his play *A la diestra de Dios Padre* (1960), turned his attention to collective creations and formed the TEC around the time the Absurd was gaining popularity.

Utopian projects, such as those seen in the New Popular Theatre movements, attempted to renovate social and cultural production and its objectives. The Cuban Revolution has been seen as the potential spark of many revolutionary projects across Latin America—both in the arts and outside—throughout the 1960s and 1970s.[18] One example of this utopian model is Augusto Boal and his *Theatre of the Oppressed* (1974), perhaps the most pertinent theorization on theater in Latin America of its time. Boal aimed to transform the role of the spectator into one that is actively involved in the spectacle. For Boal, all theater is political.[19] The theater, the physical as well as its theoretical space, is where the community can debate fundamental topics for social change. Theater is the revolutionary site, the place from which activist movements engage in dialogue with the community. Even more, as Boal noted, although theater had been used as a tool for the dominant classes, "theater is a weapon. A very efficient weapon."[20] Boal proposed that the liberating future of theater lay in the total collaboration between both sides of the stage: "First, the barrier between actors and spectators is destroyed: all must act, all must be protagonists in the necessary transformations of society."[21]

The implications of Boal's and others' theories on theater and the community were enormous in both the 1960s and 1970s given the influence that Boal had on Latin American theater even before his writings were published. Attending a play assumed a level of complicity where the spectator directly participated in a play's representation. For Boal, the differences in roles between actor and spectator are erased in order to be transformed into a new model: one where a theater is in direct communication with the surrounding community through its topics and messages. This idea of collaboration with the audience had been previously elaborated by Antonin Artaud. In "No more masterpieces" (1938), Artaud focuses on the importance of the spectator for a dramatic representation: "the spectator is in the center and the spectacle surrounds him."[22] In Theater of Cruelty, the spectator leaves behind the role of *voyeur* and becomes the center—and therefore, an essential part—of the spectacle. This step opens new possibilities, given that its reach can be multiplied by the number of people in the audience that participate and are engaged in the action—in fact, for Artaud, every spectator becomes another actor and educator. This level of involvement in the theatrical representation is what will influence the spectator's later actions. The theatrical representation, for Artaud, has the power to influence the spectator, to educate her/him to a new way of thinking and being. These concepts are fundamental to understanding the theater of the 1960s and 1970s because they reveal the power that was seen to reside in the theatrical spectacle.

In this same second half of the twentieth century, Latin American theater saw an increased emergence of women writing and publishing. Because of the inherent difficulties of writing and producing theater, women had not formed a strong part of the play writing community in the nineteenth century and the early twentieth. Writing itself, as we know from Virginia Woolf and others, is a taxing task that was not always within women's reach. Add to that, the need to have access to money to stage plays and an entry into the theater community as playwrights and the numbers can be understandably low, a similar phenomenon to cinema and women directors and screenwriters. Nevertheless, this is not to say that there were no women writing in the twentieth century and before. Perhaps the most successful and well-known among women playwrights of Latin America is

Griselda Gambaro, who entered the theater community in the 1960s, writing her first play *El campo* in 1965. Her play *Información para extranjeros* (1973) will be analyzed here in the fourth chapter.

Gambaro's is not the first or the only name that forms a part of the pantheon of women playwrights, though she is often the only woman included in anthologies of Latin American theater. Elena Garro of Mexico, in addition to narrative, was also a well-known playwright, producing and writing plays for the theater group *Poesía en Voz Alta*. Many of her plays appeared in the late 1950s and 1960s and a number can be found in the anthology *Un hogar sólido* (1958). Earlier still, Aflonsina Storni (1892–1938) of Argentina was writing theatrical farces in the 1920s and 1930s. And, undoubtedly there are many other women who were writing and contributing to the stage whose names have been lost or are unknown at this time.

Though there were not high numbers of women contributing to the play writing aspect of Spanish American theater in the early twentieth century, this has rectified itself somewhat in the following years. Perhaps one of the most recognized names is that of Rosario Castellanos (1925–1974). Her *El eterno femenino* (1973) explored the definitions of what it meant to be a Mexican woman. This play portrays a woman getting her hair done on her wedding day and looks at the various traditional roles open to women at the time. In the second act, she examines the depiction of central women in global and Mexican history, such as Eve and Sor Juana Inés de la Cruz, and how these characterizations have been damaging to women.

Myrna Casas (b. 1934) of Puerto Rico is another woman who had been able to successfully contribute to the theater of her country and region despite the stated difficulties. She began writing for the theater group *Areyto*, founded by Emilio Belaval (1903–1972), where her plays were produced but not published. Her first play *Cristal roto en el tiempo* (1960) was published as part of the Festival de Teatro of the Instituto de Cultura Puertorriqueño. Her work shows a diversity of form and topic, but she brings a perspective of women's roles to the theater.

The richness and diversity of the contributions of Latin American women to theater can be further seen in *Holy Terrors*, edited by Diana Taylor and Roselyn Costantino. This book is a collection of essays and plays written by or about the contributions of Latin American women to theater within the last thirty to forty years. Here, the editors highlight the multiple ways in which women have been able to insert themselves onto the stage, both by writing and acting and by forming theater collectives apart from some of the men in the theater.[23] As Taylor and Costantino point out in "Unimagined Communities" in *Holy Terrors*, these women, often activists, had to fight not just the outside world for a space within the theatrical communities, but also the chauvinism of men alongside whom they worked.[24] However, Taylor and Costantino suggest that women born in the 1950s and 1960s were able to enter more and more into the theater world given the increased political mobilization of women that characterized the 1970s and beyond. In this way, we see an increased number of women playwrights associated with Latin American and international stages. While Gambaro continues to write and be produced, she has been joined in the theaters and in other less traditional spaces by other names, such as Diana Raznovich in Argentina, Sabina Berman and Jesusa Rodríguez both of Mexico, Diamela Eltit of Chile, Tania Bruguera of Cuba, among many others. Nevertheless, these women and others still attest to a certain amount of resistance from different sides that make it difficult to do their work.[25]

Though the influencing factors on and of Spanish American theater may be seen to stop at the Río Grande, it is important to look farther north in the United States where Luis Valdez created Teatro Campesino in 1965. Valdez, known for such theater hits as *Zoot Suit* and *La Bamba*, asked for support from César Chávez's developing labor union to form a theater collective that would entertain and educate, as Jorge Huerta states in *Chicano Theater: Themes and Forms*.[26] This group became the beginning of a network of Chicano theaters that formed across the United States and whose influence reached beyond these national borders. Both Teatro Campesino and the network of theaters that it generated dedicated themselves to portraying issues important to the communities in which and for whom they performed, messages that were sometimes more important than the style in which they were portrayed. As Huerta affirms, "Chicano theater was born of and remains a people's theater."[27] This can be seen in the theaters' use of language, often not exclusively English or Spanish, but both, highlighting the complexities of the communities' history and the combination of recent arrivals and those who had been in the United States for centuries. While the political borders that separate the United States from Mexico may seem to eliminate discussion of Teatro Campesino and other Chicano theaters from a book dedicated to Spanish American theater, there is much more that is shared and facilitates a discussion of these collectives here, including their themes. Furthermore, Teatro Campesino both influenced and was influenced by the theater that was being written and produced on the other side of the border.

As we have seen, the theater of Spanish America has often been influenced by indigenous American factors and by European traditions to form a new theater. This theater has been connected and has contributed to the communities from which it emerged. This connection will be analyzed in the chapters to follow where this book will examine how theater aimed to create a dialogue on the change and violence that was taking place *off* stage through the use of violence *on* stage. The theater, both what was produced and written, highlighted the role of spectacle and theatricality in the political and social realms in order to urge its communities to end the use of violence.

Acknowledgments

First, I would like to thank José Quiroga for all the insight, help, and advice that he provided to me along the way. It really would not have been possible without his opinions and thoughts. I am truly honored to have worked with him and to continue to call him my mentor. I thank my editors at Palgrave Macmillan: Julia Cohen, Samantha Hasey, and Luba Ostashevsky. Working with them has been a privilege and a pleasure. Thank you to Priscilla Meléndez, whose early insight into two of the chapters helped to push the project farther and to an anonymous reader who encouraged me to think of the book more completely and globally. Thank you to Karen Stolley, who read and offered advice on the introduction. In addition, thank you is due to those who read this project in a very early version and pushed me to think more in depth to create the product it is here: John Ammerman, for agreeing to read and comment without knowing me or what the task would entail; Hernán Feldman, for reading with such interest and commitment; and, of course, María Mercedes Carrión, for inspiring this love of theater and supporting my commitment to feminist theory throughout my years at Emory and continuing to listen and challenge me. Thank you all.

I would also like to thank my colleagues at East Carolina University. First, I thank the Division of Research and Graduate Studies for start up funds that allowed me to conduct additional research in Havana and Buenos Aires. Thank you to the librarians and inter-library loan department of ECU, especially Mark Sanders and Lynda Werdal. To all of my colleagues in the Department of Foreign Languages and Literatures, thank you for understanding, for offering your insights, and for lending your ears. Your help and knowledge has been immeasurable. Thanks most especially to Charles Fantazzi, Elena Murenina, Frank Romer, and Peter Standish for reading versions of these chapters and helping with the process. And thank you to my Special Topics class on Spanish American theater—our conversations proved invaluable to many parts of this book. Also, thank you to a Library Travel Grant sponsored by the University of

Florida to their Latin American Collection that facilitated countless documents and afforded me the time to read and write in June of 2008.

Thank you to my professors and friends at Emory, Middlebury, and Bowdoin for encouraging and challenging me. There are too many that have touched my academic and personal life to count, but most especially, thank you to Eduardo Camacho, Aileen Dever, Hazel Gold, Francisco Layna, Mark Sanders, Karen Stolley, Donald Tuten, and Enrique Yepes. And to my colleagues with whom I went through the rites of passage: thank you for your support, encouragement and conversation, especially Anjela Cannarelli-Peck, Julia Carroll, George Thomas, and Scott Weintraub.

Finally, I thank all my family and friends who have supported me along the way without whom I would have been completely lost. A special thanks goes to my family: to my mother for encouraging me unconditionally; to my sister for enforcing deadlines; to my father and my brothers for their constant support; and to Rufino for always suggesting I work just a little more. I cannot measure what you have done for me—thank you.

Introduction

Difficult Times: Considering Dramatic Violence

Le explicaron después
que toda esta donación resultaría inútil
sin entregar la lengua,
porque en tiempos difíciles
nada es tan útil para atajar el odio o la mentira.

"En tiempos difíciles," Fuera del juego, Heberto Padilla

Violence, along with questions regarding national identity and political change, has defined Latin American societies throughout their history, and theater offers a way to understand how this violence has shaped the social and political contexts of Latin American societies in the twentieth century. This book examines how violence has been used in four Cuban and Argentine plays written in the late 1960s and early 1970s as a way to understand the social and political context in which they were written and performed. The plays I focus on explore the violence that appears in the theatrical context of Cuba and Argentina, two of the Latin American countries with the largest theater communities during the period from 1968 to 1974. Cuba and Argentina can be seen as a logical comparison in that they share various points of contact during the 1960s and 1970s, both in the political context and in the theater communities where individuals from one country often traveled or lived in the other. Che Guevara, of course, was an Argentine who fought alongside Fidel and Raúl Castro and was an early architect of the Cuban Revolution and the Cuban playwright Virgilio Piñera lived and worked for many years in Buenos Aires. This time period, one known globally for its turmoil, in Cuba and Argentina was a moment of change that was understood most often through violence. Though each play examined in this book approaches violence in a different way, they all represent it as a way to engage their audiences with the social and political contexts surrounding them and allow us to understand the larger framework of both these two countries and the surrounding historical and social contexts. By taking control of the violence that was being manipulated offstage to provoke fear and paralysis, the playwrights and the theater community as a whole aimed to unmask these machinations and empower their audiences. This analysis examines how theater endeavored to initiate a dialogue with its communities about the surrounding violence and how this dialogue could help bring about answers to this violence.

The Cuban and Argentine theater of the 1960s and 1970s endeavored to highlight the role of the audience in the production of violence *off* stage by calling attention to this violence *on* stage. This phenomenon will be examined through four plays, two Cuban: Virgilio Piñera's *Dos viejos pánicos* (1968) and Abelardo Estorino's *La dolorosa historia del amor secreto de don José Jacinto Milanés* (1974), and two Argentine: Eduardo Pavlovsky's *La mueca* (1971) and Griselda Gambaro's *Información para extranjeros* (1973). The purpose of this study is to examine the function of violence onstage as protest, and to understand how it was used to bring about the utopian desires characteristic of the period. The plays studied here liken the spectacle of violence that was taking place offstage with that which they were presenting onstage.

The 1960s and 1970s were decades when national definitions were being solidified against a backdrop of spectacularity.[1] That is, the political characterizations that a nation constructed officially and unofficially from within its borders and the perception from outside were being clarified through the use of spectacle. This idea of spectacle is reproduced in the four plays studied here, and it challenges the political and national projects going on offstage. To understand the goals of these plays, it is important to consider what was happening in the historical and social context of the time period to understand the close connection between political events and the theatrical stage. To that end, I will take a close look at the 1971 *caso Padilla* (Padilla Affair). Though this affair unfolded in Cuba, it reached much farther, since Cuba was seen as a utopian model for leftists throughout the Americas. Coupled with this was the fact that the affair quickly grew to be bigger than a national event when Padilla's incarceration was condemned internationally. Highlighting the spectacularity of the moment, this event can itself be seen as a constructed spectacle and, in this way, will be essential for these reflections on the connection between the Latin American stage and the political violence of the contemporary moment. This introduction will examine the construction of a political and social spectacle around the Padilla Affair and how that connects to Latin American theater of the late 1960s and early 1970s. Together with an exploration of contemporary theater and violence studies, an analysis of the Padilla Affair will allow us to understand the role of spectacle and performance in Latin America in the 1960s and 1970s and the connection that can be seen between politics and theater.

Piñera, Estorino, Pavlovsky, and Gambaro use violence onstage as a way to connect with their audience, as a way to create a dialogue between both sides of the stage that would reveal and uncover the spectacle of real violence outside the theater. This could only be done with an audience that connected with the material onstage, an innovative process of spectatorship that had begun much earlier. Latin American theater's efforts to create active and engaged spectators owe much to Antonin Artaud and Bertolt Brecht's influence, as outlined in the preface. In their work, the audience becomes a dynamic member of the representation through its reactions. In Latin America, in the 1960s, the Brazilian Augusto Boal further revolutionized this connection to the spectator with his *Teatro del oprimido* (1974). Here, Boal created a theater that drew from Paulo Freire's ideas in his *Pedagogia do oprimido* (written in Portuguese in 1968, published in Spanish and English in 1970). Boal advocated a theater that urged the actor and the spectator to join together to mutually transform the surrounding community. In Boal, theater, then, was seen as a weapon that would bring about the utopian ideals of the moment.

Understanding the role of collaboration is fundamental for any study on theater, given that the role of social change through collaboration on stage and off is central. The reaction and involvement of the spectator within a play becomes fundamental for its message to be conveyed completely; that is, through the staging of social change and violence in Latin American theater, the theater itself becomes a space of dialogue about the social process that begins to bring about protest or to cure its community.

This link to immediacy between the theater and its audience that Boal champions is also present in the writings of Bertolt Brecht on theater. As Brecht states in a short essay entitled "Can the Present-Day World be Reproduced by Means of Theatre?," "it may be enough if I anyway report my opinion that the present-day world can be reproduced even in the theatre, but only if it is understood as being capable of transformation."[2] For Brecht, the connection between the immediate context and theater is one that is centered on its ability to renovate and rework its context.

The innovative ideas of a theater in the vanguard of its community, and also initiating a dialogue with the future formed part of the contemporary Latin American framework echoed in the work of these four playwrights and was also conveyed in Che Guevara's "El socialismo y el hombre en Cuba." In this letter, written in March of 1965 to Carlos Quijano of the weekly paper *Marcha* in Montevideo, Che discussed the construction of socialism and the role of education in creating this new man for the twenty-first century. He stated that Cuba could bring about the desired socialist society and identified this as the task of the intellectual, revolutionary avant-garde: "Los hombres del partido deben tomar esa tarea entre las manos y buscar el logro del objetivo principal: educar al pueblo [The men of the party must take that task in hand and look for the achievement of the principal objective: educating the people]."[3] The avant-garde that Che recognized as educators of the people is similar to the avant-garde that Boal identified in theater, and would later write about in *Teatro del oprimido*. For Boal, theater is the space in which the revolutionary avant-garde will educate the people of future generations to bring about the dream of the utopian state, which should always maintain dialogue with "la masa [the masses]" in the words of Che.[4] Indeed, Vicente Revuelta, a well-known member of the theatrical community in Havana in the 1960s and the director of Teatro Estudio stated in an interview in 1964, that "la tarea fundamental del teatro no es la agitación, sino la contribución a la educación del pueblo para la construcción del socialismo [the fundamental task of theater is not agitation, but the contribution to the education of the people for the construction of socialism],"[5] making clear the social connection between the arts and politics. However, this is a connection that would change in the second half of the 1960s, as we see with the Padilla Affair.

One of the most interesting qualities of violence is the way that it has been integrated into society and the way that society has found to manage and manipulate violence. In *Discipline and Punish* (1975), Michel Foucault notes that in the eighteenth and nineteenth centuries there was a turn away from torture as spectacle. Before, public violence had been used as censure and a sort of entertainment that simultaneously served as a deterrent. For Foucault, the violence inherent in torture was either part of a spectacle or the very spectacle itself and something that, by the eighteenth century, needed to be hidden from public view.[6] In the modern rituals of execution, Foucault notes, the ideas of a spectacle and of pain make way for the concept of the soul and justice: "At the beginning of the nineteenth century, then, the great spectacle of physical punishment disappeared; the tortured body was avoided; the theatrical representation of pain was excluded from punishment."[7]

Foucault's comments are a particularly interesting starting point for this project because they focus on the creation of violence as spectacle. It is important to consider the use of simulated violence as dramatic spectacle in the 1960s and 1970s—150 years after the decline of public torture and execution as noted by Foucault. Why do Latin American playwrights turn to the use of violent acts to engage with their spectators and public? Why does Estorino return to the past violence of slavery to talk about censorship? Why does Pavlovsky break apart a marriage to examine the violence of relationships? For René Girard, violence changed throughout the years—it held different meanings in different times, and depending on the social contexts in which it was found. In *Violence and the Sacred*, Girard recognizes the role of sacrifice in keeping violent acts and their consequences in check: "The sacrifice serves to protect the entire community from *its own* violence."[8] In this way, violence is a central part of any community and, as such, it must be controlled: "if left unappeased, violence will accumulate until it overflows its confines and floods the surrounding area. The role of sacrifice is to stem this rising tide of indiscriminate substitutions and redirect violence into 'proper' channels."[9] For Girard, violence is an unavoidable element of all communities that is controlled through ritual. Violent acts are a natural consequence of group living, and a sacrificial bloodletting is a controlled process that prevents an all-out bloody war. Girard understands the circularity that is inherent in this process: "Only violence can put an end to violence, and that is why violence is self-propagating."[10]

For Foucault and Girard, public violence serves multiple goals within each community that cannot be replaced or substituted. Violence is in part a spectacle that serves to curb and contain a greater violence that would threaten to annihilate the community as a whole. In this way, watching a violent act can serve as a catharsis that would ideally prevent more violent acts. For Girard, simulating a violent act onstage may restrain the need to reproduce this same violent act offstage. Like Aristotle's catharsis, a representation of violence purges the spectator's need for real violence. Thus, the spectacle of violence becomes a deterrent to a real, offstage violence.

For Hannah Arendt and Frantz Fanon—two theorists of violence traditionally defined as the pacifist and the revolutionary voice respectively—violence is an inevitable part of an increasingly more deadly and destructive world. However, they differ in their interpretations of violence. For Arendt, in her study *On Violence* (1970), violence is a difficult topic to define in a time when it is being used to deter a war rather than settle an argument. Writing in the context of the Cold War, Arendt noted that violence became the threat of what could come to pass, being used to *prevent* rather than fight a war. This is an important distinction to note because war had become so deadly that to engage in it as had been done previously meant mutual annihilation. The use of violence, then, demonstrates the *failure* of power and becomes instead a surrogate for a real power: "Politically speaking, the point is that loss of power becomes a temptation to substitute violence for power [...] and that violence itself results in impotence."[11] Arendt is thus arguing *against* the use of violence, seeing it as a destructive force without benefits.[12] For Fanon, violence is the only means by which decolonization can be achieved. The colonized world is one of "violence in its natural state, and it will only yield when confronted with greater violence."[13] In Fanon's thoughts, violence stands as the only action by means of which a colonized people can defeat the yoke of its colonizers. For both Arendt and Fanon, violence defines twentieth-century history—violence is used to manipulate and change. However, they differ in their perceived benefits: Fanon sees violence to be the only way to overthrow the violent colonized world, while Arendt sees violence as a force that will change the community and world in which it is exercised to a more violent one in which violence itself defines all actions.

As an unavoidable consequence of violence, pain needs to be understood within this context. Moving from the abstraction of violence to the physicality of the body, Elaine Scarry's *The Body in Pain* considers, in part, the inexpressibility of pain: "Whatever pain achieves, it achieves in part through its unsharability, through its resistance to language."[14] Pain and the level of pain that the victim is undergoing cannot be expressed and, therefore, cannot be understood by people outside the circle of pain: "To have pain is to have *certainty*; to hear about pain is to have *doubt*."[15] The implications of pain's verbal inexpressibility are enormous for violence, given that pain can never be adequately measured from the outside. This means that the pain that violent acts impose can be questioned as false, as imagined. Therefore pain, both physical and psychological,—the primary by-product of violence for the victim—will never be understood by others and will cause its very existence to be placed in doubt. The inexpressibility of pain plays an important role within the dramatic representation of violence given that the connection between the nonverbal and the verbal is central to theater.

Elaine Scarry identifies in pain a resistance to words that connects it to the theater's use of words and gestures. Theater offers the ability to present the marriage between words and images that is essential to express pain and violence. Just as pain cannot be expressed completely through verbal communication, theater goes beyond words to use body language and gestures to connote emotions and ideas. As Scarry maintains, the victim cannot effectively tell the reader about pain in such a way that can be completely comprehended. Similarly, verbal communication is not enough in the theater. Instead, a play must *show* what is happening by using visual communication such as body language, scenery, lighting. In this way, violence and the pain that it inflicts can most fully be communicated on the stage where collaboration between the director, the actor, the playwright can begin to express this incommunicable feeling.

The infliction of pain can be seen in scenes of torture, which are often divided by sexual gender. Acts of violence are frequently gendered, and gender is constructed by means of violence. It is notable that in many plays that represent violent acts—and in all the plays studied here—the characters are divided along gender lines. That is, the torturer places him/herself in a masculine space from which he/she dominates the other while the tortured is constructed along feminine lines and is dominated by the other.[16] Even where there is an inversion of roles—where a woman becomes the torturer or a man the tortured, the torturer is always envisioned within a masculine paradigm and the tortured a feminine one.[17] This transvestism of gendered space that emerges around the use of violent acts presents a view of the construction of torture around gendered roles of power. This suggests a complexity in the use of violence in order for the individual to seize power and to occupy the space that befits it that will be investigated in this study. The importance in many Latin American communities of the role of this sexual torture is evident by its repeated presence in plays. Torture, then, is divided by sexual gender and sexual intercourse becomes a tool of power of the strong against the weak.

As can be seen in these four plays, violence was a prevalent topic in the late 1960s and early 1970s. Social unrest and violent protests characterized the news that came from places as diverse as Vietnam, France, and the United States. Violence was a central part of society in which all sides engaged, creating a vicious cycle of oppressive violence where the only perceptible way to end this oppression was through the employment of more violence—an observation of which Latin American playwrights in the 1960s and 1970s were well aware. This cycle of violence is a central issue in the four plays studied here, whose characters can only break free from the violent acts through the use of more violent gestures.

Perhaps the most important violence that these plays from 1968 to 1974 share is their relationship to official or unofficial censorship. While for Piñera, Estorino, Pavlovsky, and Gambaro, theater is the means that allowed them to reach their audience in a more direct manner, only Pavlovsky's *La mueca* was produced in the period in which it was written. While these other plays were lauded officially or unofficially and were undoubtedly recognized as products of the most important playwrights of the moment, they were not produced at the time. Yet, their importance in the theatrical production of the historical and social moment cannot be questioned since they were publicly read and discussed in theater circles, even if they were not staged then. There was a tradition of reading theater publicly, so that, although a play may not have premiered, there was often a public presentation of it. These practices show that, though these plays were not performed publicly, they were in the public eye and influenced the theater communities and their audiences.[18] The difficulty of production that can be seen around these four plays, however, brings to the forefront of the discussion the violence of censorship.

This interplay between a theatrical text and its physical representation makes a major contribution to the play's success and permanence in national canons. Theater is a genre that, like cinema, is much more susceptible to unofficial censorship, given its need for both money and labor to be staged. The story of these plays begins with the violence done to prevent their productions. A play can be awarded a prestigious award but not be produced within the temporal context in which it was written (Piñera's *Dos viejos pánicos*) or a play alluding to the national context of fear and violence of the time can be written but not produced or even published for several years, circulating unofficially both within and outside the country (Gambaro *Información para extranjeros*). In contrast, a play can enjoy a timely premiere thanks in part to the acting abilities of its author (Pavlovsky's *La mueca*) or appear to be on the verge of a premiere that is pushed farther and farther back (Estorino's *La dolorosa historia*). The various levels of censorship that can be seen in these four plays places on center stage the violence that is inherent in this social and political act.

However, to say that censored plays never enter into the public sphere would be a mistake in that many of them circulated in the theater community and were read by different members of that community and the public. For example, *Dos viejos pánicos* did not premiere in Cuba until many years after it won the Casa de las Américas prize, but this prize guaranteed that it would be discussed and read by those in literary and theater circles. Similarly, Gambaro's *Información para extranjeros* did not premiere and was not published in Argentina until much later, but it was read clandestinely in the country and openly circulated abroad. In this way, though the plays do not reach the full fruition that is first imagined when they are written, they do initiate a dialogue that others in the community are able to carry on and consider.

In Piñera, Estorino, Gambaro, and Pavlovsky, violence becomes a spectacle used to alert the audience to wrongdoings happening offstage. The dramatic representation of violence creates a space from which the playwrights and the theater community can enter into a dialogue with their spectators and public on the issue of violence and its place in their communities. These plays show how theater offers the ability for both the actors and the spectators to engage with one another and to collaborate on an appropriate reaction to official violence and what it will mean for the community. This conversation is meant, then, to deter an offstage violence and create a discussion on violence that enters into debate with the historical and political moment in Latin America in the 1960s and 1970s. This discussion of the topic of violence is not new or particular to the theater. It has been addressed by a number of recent books on modern Latin American literature, including Susana Rotker's *Ciudadanías del miedo* (2000), Rebecca Biron's *Murder and Masculinity: Violent Fictions of Twentieth Century Latin America* (2000) and Aníbal González's *Killer Books: Writing, Violence, and Ethics in Modern Spanish American Narrative* (2001), among others. Rotker's collected essays offer a sociological view of how fear and violence have constructed a people, while Biron and González examined how fiction has reflected on or, conversely, erected this omnipresent violence within Latin America. These writers all address different aspects of violence and have separate viewpoints on its manifestation, but the continued presence of the topic of violence in the twentieth century points to an intense desire to delve deeper into this question. In theater studies, Severino João Albuquerque's *Violent Acts: A Study of Contemporary Latin American Theatre* (1991) focuses on this connection between theater and violence. Albuquerque provides an impressive overview on the use of violence in theater. In turn, Amalia Gladhart's *The Leper in Blue: Coercive Performance and the Contemporary Latin American Theater* (2000) examines the potential danger of the performance space for both the performer and the spectator. Gladhart's is an impressive study on the coercion that is involved in performance and how this is used in Latin American theater.

Idelber Avelar's *The Letter of Violence: Essays on Narrative, Ethics, and Politics* (2004) focuses its attention on the role of violence in modern society with the aid of various reflections on violence from Continental critical theory, Anglo-American philosophy and Latin American literature, such as Carl von Clausewitz, Michel Foucault, and Paul Virilio. Avelar recognizes the centrality of violence in the construction of a modern state: "Going back to Thomas Hobbes's concept of a state of war as the originary, enabling figure for all history, the problem of violence has occupied a key position in modern Western thought."[19] Violence is the central piece needed to understand the workings of modern society, whether or not the subject in question exists within the space of a violent act. It is to this growing body of academic criticism that this book seeks to contribute.

Even more recently, Hermann Herlinghaus has considered the intersection of violence and guilt in *Violence without Guilt: Ethical Narratives from the Global South* (2009). Herlinghaus uses Walter Benjamin's writings on violence to examine the prevalence of violence in the Americas due to the increased war on drugs that marked the 1980s and beyond. He focuses particularly on the border area between Mexico and the United States and on Colombia, both focal points of drug violence in the twentieth century.

Violence without Guilt goes beyond many earlier studies on violence by focusing on the *narcocorrido* in Mexico and the use of film in Latin America. Given this widened view and his concentration on the 1970s and beyond, *Politics and Violence in Cuban and Argentine Theater* serves as a necessary complement that examines the genre of theater in the 1960s and 1970s. Herlinghaus proves that what will be discussed in reference to Cuba and Argentina in the late 1960s and early 1970s was not contained within these years or particular to these countries, but defied geographical and temporal borders. The drug violence that Herlinghaus discusses, though, as examined in this book, is not the first violent intervention by governments or extra-official forces. Violence has always been a part of all times and contexts, a fact that we can see in the 1960s and 1970s.

It is essential to note that violence defined the late 1960s and early 1970s in a way that has marked the world since. Given the various global and local events—the Vietnam War, the student protests in places like Paris and Mexico City, the assassination of Martin Luther King, Jr. and Robert Kennedy, the Cuban Revolution, to name a few—violence and the threat of future violence loomed over everyday activities as an imminent possibility. At this time, violence appearing onstage was used as a tool that could question or protest a simultaneous violence in the immediate context. Many playwrights and artists, such as Augusto Boal of Brazil and Osvaldo Dragún of Argentina, believed that the arts could be a place of dialogue and innovation where the artist and his/her public could search for a way to achieve their utopian desires together. Theater was the community activity that aimed to engage its audience in a dialogue about its future and, thus, renovate the social environment in which these plays were written.

The twentieth century in general is a time known for extreme barbarity. The historian Eric Hobsbawm names the period between 1914 and 1991 "the Short Twentieth Century" where everything "is marked by war," a fact that is fundamental to understanding this century.[20] The years between the two World Wars were defined by a global absorption with events that would culminate with the Second World War, thus unifying the years leading up to and including both Wars. Although the Latin American republics were only marginally involved,[21] the brutality of war in Europe profoundly affected the entire world, and Hobsbawm underlines the political consequences of this bloody destruction: "The experience itself naturally helped to brutalize both warfare and politics: if one could be conducted without counting the human or any other costs, why not the other?"[22] Hobsbawm's generally European focus notwithstanding, it is important to remember the fundamental place that that continent occupied in the world, and above all, in Latin America, given the recent past of colonialism and the commercial activity. In this way, the violence and brutality that marked European history in Hobsbawm's view of the twentieth century sees its effects mirrored in Latin America, where it was a time marked by bloody revolutions like the Cuban one and violent oppressions like those in Argentina.

The role of political commitment and utopian projects in this historical time frame is defined in many ways by questions of violence, politics and social change. That is, artists were measured by their perceived allegiance to political ideals, ones that were defined by a violent need for social change. *Dos viejos pánicos, La dolorosa historia del amor secreto de don José Jacinto Milanés, La mueca*, and *Información para extranjeros* are written within, and contribute to, a highly politicized context that judged them on their perceived intensity of commitment to the political situation. The triumph of the Cuban Revolution in 1959 with Fidel Castro's capture of Havana ignited a fervor throughout Latin America that opened the possibility of socialism for the continent and inspired many revolutionary actions and consequent repressive reactions by many established governments. These possibilities can be observed in Che Guevara's travels to spread the revolution to Africa in 1965 and Bolivia in 1967, while his untimely death in 1967 in Bolivia demonstrates the violent reactions of these acts. In reactionary responses to the growing wave of socialist and Marxist insurrections, some governments turned their attention to their own citizens in order to control the spread of the new ideologies, with disastrous results. An example of this repression can be seen in Mexico in the Massacre of 1968 (the year *Dos viejos pánicos* was written), when Mexican President Gustavo Díaz Ordaz ordered the massacre of protesters, many of them students, in the Tlatelolco Plaza of Mexico City. In 1970, when Salvador Allende was elected president of Chile, the possibilities of democratically elected change were offered as an alternative to the Cuban model. But these possibilities were shattered in 1973 when the Chilean Way to Socialism, and Allende's life, ended in La Moneda.

Perhaps the most pivotal of these events is the Padilla Affair, given the spectacularity and staging that was central to its execution.[23] The Padilla Affair is an appropriate example of the creation of a public spectacle meant to inspire fear and complicity in the audience, in other words, the public. While this episode takes place in Cuba, it is without a doubt an international affair that sparked intervention from well-known European and Latin American intellectuals and signaled a turn to suppression and spectacle within the Cuban Revolution. The audience of the events was firstly the Cuban people, but the entire world was captivated by what was unfolding and the effects were felt beyond the Cuban borders. What has come to be known in Spanish as the *Caso Padilla* (Padilla Affair) involved the 1968 awarding of the theater and poetry literary prizes of the *Unión de Escritores y Artistas de Cuba* (Union of Writers and Artists of Cuba [UNEAC]) for Antón Arrufat's play *Los siete contra Tebas* and Herberto Padilla's collection of poems *Fuera del juego*. The UNEAC was an important group that was created after the Revolution, to support writers and artists; the annual prizes were given in the areas of narrative, poetry and theater. These two works received the prestigious awards despite their supposed "counter-revolutionary" ideology, as the UNEAC stated in a declaration that was printed with and about both of the works. Both authors suffered censorship and alienation from Havana's intellectual circles. This affair became an international episode that sparked spectacular breaks and criticisms of all sides. Heberto Padilla (1932–2000), at this time, was already considered an important poet. Like many others, he had returned to Havana from exile with the triumph of the Revolution and found work connected to the new governmental organizations created for the arts. By the middle of the 1960s, however, like others in the literary community, he was beginning to question the Cuban government's official path. It is helpful to understand the construction of this spectacle in order to view the dialogue between theater and politics that marked the late 1960s and early 1970s. The Padilla Affair offers an invaluable example that allows us to explore this connection.

Though the award is perhaps the most famous event, it is actually the midpoint of the affair. In 1967 Padilla published an essay in *El caimán barbudo* that supported Guillermo Cabrera Infante's *Tres tristes tigres* over Lisandro Otero's *Pasión de Urbino*. Both novels had been finalists in the Concurso Biblioteca Breve of the Barcelona-based Editorial Seix Barral in 1964, though Cabrera Infante's won.[24] By this time, Cabrera Infante had broken with the Revolution and was living in exile in London. The fact that his novel was awarded instead of the revolutionary Otero's was a point of contention that Otero, the Vice-President of the National Cultural Council, and others decided to protest. Padilla, however, did not go along with this protest and instead took the opportunity to publicly support Cabrera Infante's novel and to criticize the UNEAC.[25]

This gesture was answered with official criticism of Padilla and the dismissal of the entire editorial staff of *El caimán barbudo*. Padilla, though, continued writing, publishing another critical essay in *El caimán barbudo* in March of 1968. This same year he also submitted his unpublished collection of poetry, titled *Fuera del juego* [*Out of the game*] to the UNEAC's Julián del Casal prize for poetry. *Fuera del juego* was considered the clearly superior choice for the prize. The jury, composed of the Cubans Manuel Díaz Martínez, José Lezama Lima, José Z. Tallet, the Peruvian César Calvo, and J.M. Cohen of Great Britain, was pressured to give this award to another.[26] They held out, however, and Padilla's collection won, though it and Antón Arrufat's *Los siete contra Tebas* were published with a letter denouncing them: "son ideológicamente contrarios a nuestra Revolución [they are ideologically against our Revolution]."[27]

This critique, in turn, inspired more official criticism both of and from Padilla, culminating with another collection of poetry, intriguingly titled *Provocaciones*. When he publicly read some of these poems, the state arrested him and his wife, the poet Belkis Cuza Malé on March 20, 1971 and searched his house, finding five copies of his unpublished novel *En mi jardín pastan los heroes*, but accidentally leaving the original which had been hidden in a basket of toys. This move sparked international criticism of the Revolution by such earlier supporters as Octavio Paz, Gabriel García Márquez, Simone de Beauvoir and Jean-Paul Sartre, among others, all of whom signed an open letter to Fidel Castro, published in *Le monde* on April 9, 1971. This was almost three weeks after Padilla had been arrested and more than two weeks before he would be released and would read his confession at the UNEAC headquarters. The letter was short—about a page—and asked Fidel Castro to re-examine the situation that had led to Padilla's arrest. The writers' of the letter urged against repressive measures citing the tenuous situations of other Latin American countries, such as the newly-elected socialist government of Chile. They said that the use of force against intellectuals and artists "puede únicamente tener repercusiones sumamente negativas entre las fuerzas anti-imperialistas del mundo entero, y muy especialmente en la América Latina, para quienes la Revolución Cubana representa un símbolo y estandarte [can only have extremely negative repercussions among the anti-imperialist forces of the entire world, and most especially in Latin America, for whom the Cuban Revolution represents a symbol and standard]."[28] They closed the letter by reaffirming their solidarity with the Revolution. In both this statement and the closing show of support, we see a conflict between an intellectual public urging Fidel to yield in this situation and a reaffirmation of the Revolution and what it stands for—all taking place in the public eye as if upon a stage.

After the two were arrested, Belkis Cuza Malé was released within a few days of detention, but Padilla remained in custody until April 27. As a condition of his release, he signed a confession which he read publicly that night at the UNEAC offices in El Vedado, Havana, in front of his colleagues. In this self-criticism, as he referred to it, he accused himself of being pessimistic and counterrevolutionary, mentioning also the names of some of his friends with whom he had discussed the Revolution.

This public self-criticism was followed by another open letter to Fidel Castro from leftist intellectuals where they publicly broke with the

Revolution.[29] The second letter from the European and Latin American intellectuals to Fidel Castro appeared in the newspaper *Madrid* on May 21, 1971 and denounced Padilla's public confession, saying that "recuerda los momentos más sórdidos de la época stalinista, sus juicios prefabricados y sus cacerías de brujas [it remembers the most sordid moments of the Stalinist epoch, its prefabricated judgments and its witch hunts]."[30] With these words the intellectuals publicly chastised the Revolution for its behavior, likening these acts to those in socialist countries during the worst of Stalinism. This letter shows that the Padilla Affair proved to be an embarrassment abroad for the Revolution. It was, however, a success within the country. This year, 1971, was the first of what has come to be known as the *quinquenio gris*, a time of censorship, both official and unofficial, and intellectual repression. This was stated officially in Fidel's closing speech at the Congress on Education and Culture, when he said that education would be the only goal of everything published in Cuba.[31] In addition, these letters allowed Castro to create another enemy of the Cuban revolution out of the "false" intellectuals: "esos que desde París ellos desprecian, porque los miran como unos aprendices, como unos pobrecitos y unos infelices que no tienen fama internacional. Y esos señores buscan la fama, aunque sea la peor fama [those who from Paris scorn because they look on them as apprentices, as poor wretches that are not internationally famous. And those men are looking for fame, even if it is the worst kind of fame]"[32] With these words, Fidel tried to dismiss his recent problems with intellectuals both within the country and outside of it, especially Padilla and his *Fuera del juego*.

What is most interesting of the Padilla episode is the construction of spectacle that is created around these events and how they contribute to an orchestrated theater of actions that allows us to understand more fully the context of the late 1960s and early 1970s. Looking at Padilla's public presentation of his self-criticism is a pivotal moment in this affair. Here, as Reinaldo Arenas has pointed out in his essay "El caso y el ocaso de Padilla," was a spectacle that was created in order to condemn Padilla's past actions and quash any future criticisms of the Revolution for the internal audience. In this way, the Padilla Affair is not about Padilla and his counterrevolutionary opinions, but a representation that warns others. It is a public service event that reaches beyond that day and those words.

Arenas, and others, sees Padilla's self-criticism as the ultimate act of irony where, in the Cuban fashion, he makes fun of the event, the Revolutionary officials in attendance, and the process that has obligated him to appear there. In this way, he is a contributor to the creation of this spectacle and begins to gain a certain amount of control over it. This irony that Arenas points out is upheld by Padilla's own words in the self-criticism. Here he speaks about the construction of an image of himself with which he fell in love. This image was embodied in a photo of Padilla by Lee Lockwood, a U.S. photojournalist, that was published with the caption "Heberto Padilla, poeta y *enfant terrible* (—niño terrible—) político [Heberto Padilla, poet and political *enfant terrible* (—terrible child—)]."[33] Padilla responded to this in his speech by saying, "En fin, me enamoré de esa imagen. Pero esa imagen ¿a dónde me llevaba? [...] ¿De qué se beneficiaba esa terribilidad sino de la enemistad con la Revolución? ¿De qué se beneficiaba sino de la contrarrevolución, del desafecto, del veneno? [In short, I fell in love with that image. But that image, where did it take me? [...] What did that terribleness benefit but the enemies of the Revolution? What did it benefit but the counterrevolution, disaffection, poison?]"[34] This reaction is particularly important in that it reveals the different levels of spectacle inherent in this act. First, Padilla states that he fell in love with an image of himself that was created in part from the outside. From there, we see that he then considered this image while under detention with the Department of Security and then takes on another role in this public self-incriminatory speech. Here there are multiple levels of representation of who Heberto Padilla is and what exactly he is saying. The irony, then, is palpable and pivotal to this construction of spectacle. In this way, Padilla gains some control over the performance in which he is acting.

As seen above, this exchange plays out in the public eye through the publication of essays in journals.[35] This control is something that can be seen to vacillate between Padilla and the Revolution throughout the events of these years. Though we have seen how Padilla was able to manipulate images or events, he was primarily at the mercy of the official organizations and groups, people that worked under the cover of pseudonyms or with the protection of the Revolutionary government. Padilla's dissent was the subject of a series of articles beginning in November of 1968 that were published in *Verde Olivo* under the pseudonym Leopoldo Ávila (around the time when the UNEAC jury was considering the award and being pressured to give it to another).[36] *Verde Olivo* was the magazine of the *Fuerzas Armadas Revolucionarias* (Revolutionary Armed Forces) in Cuba. As Lourdes Casal details in her collection of documents relating to the Padilla Affair, Ávila wrote essays about Cabrera Infante, Antón Arrufat (the 'counterrevolutionary' playwright of the 1968 UNEAC prizes) and Padilla. This last one, entitled "Las provocaciones de Padilla," (the very word that Padilla would later use for his next collection of poems) details how Padilla had changed from a promising young poet to a "viajero infatigable [tireless traveler]" who did not understand the Revolution and, by extension, its benefits to the Cuban people.[37] Padilla was painted as a provocateur who was writing for an international audience, accusing the

Revolution in his poems of censorship and other offenses, all of which, Ávila stated, the Cuban people would not believe. Besides being counter-revolutionary, in Ávila's words, these poems were "bastante malos [rather bad]" and their only reason for winning the UNEAC prize, was the fact that Padilla's friend César López formed part of the jury.

This article that confronted Padilla and his work directly was followed by another signed by Leopoldo Ávila that discussed the role of criticism in Cuban literature more broadly. "Sobre algunas corrientes de la crítica y la literatura en Cuba" was originally published in the November 24, 1968 issue of *Verde Olivo*, but received a much wider readership when it was republished in *La Gaceta de Cuba* in their November-December issue. *La Gaceta de Cuba* is the journal published by the UNEAC, and therefore one of the most important outlets of cultural production. *La Gaceta* typically announced the winners of the UNEAC prizes for literature. However, in 1968, the reader found Ávila's article on the role of criticism in literature, an interesting addition that speaks to the article itself and the events that were happening in reference to both Padilla's *Fuera del juego* and Antón Arrufat's *Los siete contra Tebas* and in the surrounding context. In his article, Ávila laments the "aparente despolitización [apparent depolitization]" of Cuban literature, despite Fidel's repeated calls to intellectuals and artists to integrate themselves into the Revolution through such acts as traveling through the provinces to see the Revolution's benefits and successes.[38]

These articles are a part of the extensive spectacle, one which Padilla again is able to direct in part when he uses a word from Ávila's very title—*Provocaciones*—for his next collection of poems, published in 1973 in Madrid, but written by 1971. Padilla refers to his experiences in Cuba of censorship and political pressure and even his public confession. The introduction details the Padilla Affair and discusses Padilla's earlier collections of poetry, ending with a discussion of the poems of *Provocaciones*. The poem "A Galileo [To Galileo]" draws a connection between the trials and the confession of Galileo by the Catholic Church and those that Padilla experienced at the hands of the Revolution. The poetic voice refers to the role of prison—la cárcel—in the creation of history: "No hay verdadera historia // que no tenga como fondo una cárcel [There is no true story // that does not have as its background a jail]."[39] Prison plays a pivotal role in this creative process, given that poetry's role here is to denounce what has happened. The poem connects the historical facts that surround Galileo with the Padilla Affair in the title and the content. This is a connection that was seen by witnesses of the self-criticism, such as Reinaldo Arenas who highlighted Padilla's connection to Galileo and to others condemned in the past (particularly Soviet dissidents) who were forced to thank the 'purifying' act that also brings about their end.[40] Roger Reed reports in

The Cultural Revolution in Cuba that Padilla said privately to some "Sin embargo, se mueve [Nevertheless, it moves]" after his 1971 confession. This is a quote from Galileo after he was forced to retract his support of Copernicus' theory that the Earth moved around the sun.[41] Reed goes on to point out the various ways that Padilla drew parallels between his own retraction and that of Galileo and some Soviet intellectuals forced to confess during Stalinist repressions. In the words of Reed, this became "a truly grotesque piece of theater."[42] We see here the creation of a public spectacle: Padilla knew his only options were retraction or continued and increased persecution. His move, then, was to create a theater of events for public consumption that would temporarily placate perhaps, but would also expose the repression of the Revolution.

To return to the collection that won the UNEAC prize, the poem "En tiempos difíciles [In Difficult Times]," which serves as epigraph to this introduction, opens Padilla's *Fuera del juego* and typifies the emotion and sacrifice that characterized the end of the 1960s and would foretell Padilla's—and many others'—fate. This poem had been published earlier in Cuba in the July–August 1968 issue of *La Gaceta de Cuba*, forming a part of the Festival of Poetry of 1968. In the poem Padilla details how a man is asked to sacrifice everything: his time, his hands, his eyes, his lips, his legs—even his tongue:

> Le explicaron después
> que toda esta donación resultaría inútil
> sin entregar la lengua,
> porque en tiempos difíciles
> nada es tan útil para atajar el odio o la mentira.
> Y finalmente le rogaron que, por favor, echase a andar
> porque en tiempos difíciles
> ésta es, sin duda, la prueba decisiva
> [They explained to him afterwards
> that all these gifts would be useless
> unless he surrendered his tongue,
> since in difficult times
> there is nothing more useful for warding off hatred and lies.
> And finally they asked him
> kindly to start walking,
> since in difficult times
> this is undoubtedly the decisive proof].[43]

With this poem, Padilla reveals what he and others have been asked to sacrifice for the future and, by not exalting this, he shows his opinion of this loss and its benefits. He does not simply toe the party line and agree that this is what needs to happen, but condemns these demands by presenting them without ceremony or grandeur. With this opening poem to the collection, there is little question that this book is not an endorsement of the revolutionary status quo. Instead, it questions this status quo in a similar way to that which Castro and the other Revolutionaries did during the 1950s and the early 1960s when they were striving to build a new future. By this time (1968), however, consolidation of what the Revolution meant was more important and this could only be done by putting forth one unyielding definition of Revolution and squashing any criticism of this definition.

This examination of the Padilla Affair is of pivotal importance to this study because it helps to understand the connection that was prevalent in the late 1960s between politics and the creation of spectacle. The Padilla Affair, starting with Padilla's assertions about the novel *Pasión de Urbino* and continuing with his *Provocaciones*, became a game where the poet criticized and the Revolution intervened in various ways—essays written under pseudonyms, an award whose merit was tempered, and finally an arrest and a public confession. It can be argued that Padilla often was able to exercise a certain level of agency over these interventions, but a close look at the Affair as a whole reveals how the Revolution attempted to stage certain reactions and responses, both internationally and nationally. Both sides, though, were vitally aware of the theatricality in their moves; they were being judged by their actions and reactions, a judgment that they wanted to control and manipulate as much as possible.

Spectacle held a central role in Latin America in the late 1960s and early 1970s, but not just on the physical stage of a theater. Instead, spectacle and spectacularity expanded its frontiers beyond the theatrical space and became an overt political tool. In this book, we will examine how four playwrights endeavored to take this use of spectacle—one that had been used repeatedly in the Padilla Affair—into their own plays in order to create a dialogue with the external world about the prevalence of violence. These plays were to open a conversation on the centrality of violence in these Latin American communities in the late 1960s and the early 1970s, through the collaboration between theater and its audiences. The spectators' connection and involvement would be key to this conversation. This creation of spectacle, or spectacularity, that formed an integral part of the time was questioned through the stage in order to highlight the silence and complicity with which public and official violence was received. This double spectacle—characters on stage (or on the page but written for the stage) simulating a spectacle that was being performed in official channels—underlined the spectacularity found throughout the community. It should be remembered that, though not all these plays premiered in their time, they all entered the public debate on the violence that was taking place through readings and partial representations. This spectacularity will be explored in this book through four plays from Cuba and Argentina.

This book is divided into four chapters that examine four different plays: violence and fear in Piñera, violence and history in Estorino, violence and identity in Pavlovsky, and violence and spectatorship in Gambaro. Chapter 1 looks at the role of fear in relation to violence in Virgilio Piñera's *Dos viejos pánicos*. In 1968 Piñera was awarded the Casa de las Américas prize (another point in common for many of these plays) for *Dos viejos pánicos*, a play that has often been considered an example of Latin America's theater of the absurd.[44] The play was published in the same year by Casa de las Américas, and it premiered in 1969 in Bogotá, though it wasn't produced in Cuba until 1990 because of the censorship of Piñera's works in the 1970s and 1980s. The play is divided into two acts with two characters, Tota and Tabo, both "viejos de sesenta años [old people about sixty],"[45] although there is a third element so pervasive and tangible in the play that one can think of it as another character: fear. The two characters spend the first act taking turns "dying" and describing their lives free from fear, while in the second act, over their daily glass of milk, they discuss life and death. Piñera is pointing to a social tendency to fear the unknown and is challenging the spectator to break through this; *Dos viejos pánicos* causes its spectators and readers to interrogate the role of fear and violence in their own relationships and the actions it leads them to take.

Dos viejos pánicos is often considered the play that officially established Piñera's centrality in the Cuban and Latin American theater community. Piñera's prominence began with *Electra Garrigó* (1941) and continued throughout his life with a particular vitality in the decade of the 1960s. Rine Leal, in the prologue to the 2002 edition of Piñera's *Teatro completo*, stated that,

> En 1968 el premio Casa de las Américas a *Dos viejos pánicos* consolidó su nombre, fue algo así como el reconocimiento oficial a su maestría y obra. Y los hombres de mi generación continuábamos viéndolo como "nuestro" dramaturgo, a pesar de los nuevos escritores, a pesar de que el teatro cubano no se había detenido en sus piezas.
>
> [In 1968 the Casa de las Américas prize to *Dos viejos pánicos* consolidated his name, it was something like the official recognition of his mastery and work. And men of my generation, we continued seeing him as "our" playwright, despite the new writers, despite the fact that Cuban theater had not focused on his plays].[46]

Given the centrality of Piñera's theater that Leal identifies here and in the prologue more generally, *Dos viejos pánicos* is the logical starting place for this study—a point of origin from which to appreciate the location of Latin American theater in the 1960s and to explore the directions in which it moved.

In Chapter 2, I turn to the use of violence in the construction of history in Abelardo Estorino's *La dolorosa historia del amor secreto de don José Jacinto Milanés* (1974). Estorino, the playwright whose *El robo del cochino* in 1961 was lauded very early in the Revolution, wrote *La dolorosa historia* in 1974, though it did not premiere until 1985 by the theater group *Teatro Irrumpe* directed by Roberto Blanco.[47] Estorino is considered one of the most influential Cuban playwrights of the Revolution, given that he often turned his attention to family and history. This play recalls the nineteenth-century Cuban poet José Jacinto Milanés (1814–1863) and the context in which he lived, particularly the large family from which he came, the impossible love he had for his cousin Isabel, and his acknowledged loss of reason towards the end of his life. Milanés is an important figure for Cuban playwrights of the Twentieth Century because of his play *El Conde Alarcos* (1838), considered the first major contribution to Cuban theater. The final scene of *La dolorosa historia*, "Delirio," which retells Milanés' loss of reason, mixes his interior life with an account of a slave rebellion in order to put race relations in nineteenth-century Cuba at the center of the stage, simultaneously creating a parallel between the private (Milanés) and the historical (slavery). The violence depicted on stage in this play is not limited to the rebellion, but alludes to the social violence inherent in a slave-owning and racist society, while Milanés' questioning of the necessity for violence in a social uprising recalls the context of the Cuban Revolution. Estorino's play finds a circularity seen throughout history and the censorship of the nineteenth century, and thus, implicitly alludes to the stronger regulations around literature in Cuba that came into effect in 1971. The discussion in the final scene of the use of violence in an uprising, and of social equity in Cuba, allows for a renewed consideration of the Cuban Revolution through the lens of nineteenth century historical events. This chapter examines the relationship between history and violence in Estorino's play in order to understand how the two interact and affect our perceptions.

The next two plays and chapters change context to Argentina in the first half of the 1970s. Chapter 3 looks at violence in relation to the creation of identity and class in Eduardo Pavlovsky's *La mueca* (1971). The aim of this chapter is to explore the relationship between violence and the construction of identity (class, gender, sexuality) in Pavlovsky's play. Class and gender were two focal points of violence in Latin America in the 1960s and a particularly important topic of cultural production. Since the 1960s, Eduardo Pavlovsky has been considered one of Argentina's most important playwrights and his influence has been felt beyond Argentine theater—most notably in Cuban theater, where he has participated in conversations on Cuban and Argentine theatrical production.

Pavlovsky's *La mueca* is divided into two acts and portrays the encounter between four men of the underworld and a bourgeois couple.[48] The men break into a middle-class couple's empty house in order to create a film about bourgeois life. The couple's return creates a clash of classes and ideologies that culminates in the destruction of façades and social hypocrisies. The men destroy the last shreds of the couple's relationship by revealing their mutual infidelities and betrayals and they uncover the bourgeois hypocrisies in which the couple takes part. The play focuses its attention on class tensions in order to displace the bourgeois couple as the conventional object of the theatrical gaze, while the physical violence inflicted by the men—who represent a lower class—suggests the omnipresence of violence in these divisions and the need for a dominant figure. Furthermore, the function of performance within the boundaries of each individual role—consciously playing a part—is a central topic that considers the defined space that each character is meant to occupy within the representation and within their community.

Finally, chapter 4 turns to spectatorship and violence in Griselda Gambaro's *Información para extranjeros* (1973) and allows us the opportunity to explore the intersections of theater and performance. Griselda Gambaro can be considered one of the most important dramatists in Latin America of the twentieth century. Her list of plays from the 1960s and beyond is extensive and central to the history of Latin American theatrical production. Gambaro's theater is often focused on the relationship between the dominated and the dominator, an idea from which she somewhat departs in *Información para extranjeros*, concentrating instead on the role that violence has assumed in her immediate national and political context. Gambaro has this debate take place within an innovative space and with an engaged spectator that questions the boundaries of theater.

These boundaries of the genre of theater can be seen to be flexible when considered alongside many other types of visual arts. In some recent scholarship, theater is often eclipsed by, or confused with, performance, a term that should be defined in relation to this play since they can be seen to connect here. Patrice Pavis, in his *Dictionary of Theatre*, emphasizes with performance "the ephemeral and unfinished nature of the production rather than a completed work of art."[49] Diana Taylor, in "Opening Remarks" to *Negotiating Performance: Gender, Sexuality, and Theatricality in Latin/o America*, stresses the importance of the actor: "Performance art tends to be based in the actor rather than the text; it tends to be personal and grounded in narrative [...] the focus of performance art is on the hermetic and private rather than the public."⁵⁰ These definitions of performance, like many others, seem oftentimes to bleed into that of theater and connect many aspects between the two genres. As we will see, Griselda Gambaro's *Información para extranjeros* plays with these definitions in order to innovate the notion of theater and engage her audience in a dialogue on the contemporary violence in Argentina, making this work an appropriate place to conclude.

Información para extranjeros is a play that leaves behind the traditional idea of a theatrical representation as one comprised of spectators in seats removed physically from the action on stage.[51] The play never premiered entirely in Argentina, though selected scenes were staged in the provinces; it has been performed in Mexico and Germany. Gambaro divides the argument into twenty scenes of unequal length that create an experience where the spectators are guided through rooms in which atrocities and disappearances are taking place. The scenes which are labeled "Información: para extranjeros [Information: For Foreigners]" by the guide have been taken from contemporary newspapers.[52] For example, in the first scene, after the guide has greeted the group, he leads them to a room and, in the process, warns "¡Cuidado con los bolsillos! [Watch your pocketbooks!]"[53] Following this warning, he opens the door to the room in order to view a man who is clothed only in a loincloth. The guide quickly shuts the door, stating "Perdonen. Me equivoqué de habitación [Excuse me. I've got the wrong room]."[54] This play forces the spectator to see what is happening outside the theater, rather than gloss over what is distressing. In that way, the violence onstage is directly linked to that which is happening offstage. This chapter will examine the connection between the spectator and violence in order to understand Gambaro's theatrical use of space and spectatorship and the implications of these for the theatrical community and political context.

The portrayal of violence within the pages, or stages, of these plays is paralleled by the unpredictability of the historical moment in which they take place and are written. The plays span six years (1968–1974) full of violent acts with far-reaching repercussions felt beyond each of their national boundaries. It is important to remember the violence outside of the theatrical production and the context in which the plays were written. Indeed, it is my assertion that the unpredictability that is seen in the violent historical events of the period has a direct relationship to the plays' difficulties of production at the time when they were written because the playwrights were writing *against* the contemporary official or standard use of violence in their communities. These playwrights' inability to intervene in the cultural and artistic production of the time owes itself, in part, to the unpredictability that is characteristic of any theatrical production, given the collaboration that defines the theater. However, this unpredictability is particularly important for these plays because of the instability of their portrayal of violence. That is, violence is used as an element that undoes and questions within these four plays. It is the project of this book to understand more fully the use and the objective of violence in these plays and the dialogue that was initiated with their spectators *and* readers.

In the 1960s and 1970s, theater was the space in which political and social debates initiated a dialogue between intellectuals and popular audiences; similarly, the context of the 1960s and 1970s is one of violence and change. Latin American theater combined this connection with the audience and its engagement with the moment to question the spectacle of violence that was taking place offstage. By presenting violent acts onstage, theater highlighted the role of silence in proliferating this violence and underlined what it meant for the country and the people. Each play examined in this study illustrates a different manifestation of what violence was and meant within the context of Cuba and Argentina in the late 1960s and early 1970s. This book explores how debates on theater take place around the topic of violence in order to understand more fully the role of violence in the theater. Piñera, Estorino, Pavlovsky, and Gambaro influenced their public and the larger theater communities to be aware of the violence surrounding them in an effort to put an end to the violence. The playwrights do not attempt to dictate actions in a didactic manner but instead want to comprehend this global violence through the stage and the collaboration inherent within theater. This can be seen as a radical act in that the plays were meant to begin a dialogue that would unmask violence and question its purpose within the communities. Each play wants to uncover how violence is used in quotidian existence in order to understand its pivotal role in relationships and human life. Piñera, Estorino, Pavlovsky, and Gambaro use the stage as a means to consider this violence, a radical act that has the potential to eradicate a wider-reaching and deadly violence, one that was a global occurrence during the 1960s and 1970s.

Chapter 1

Who's Afraid of Virgilio Piñera? Violence and Fear in *Dos viejos pánicos* (1968)

Tabo: Tota, ¿qué vamos a comer mañana?
Tota: Carne con miedo, mi amor, carne con miedo.

Virglio Piñera Dos viejos pánicos

Dos viejos pánicos (1968) from Vigilio Piñera (1912–1979) explores the violence inherent in the repetition of everyday existence by portraying a routine day in the life of a sixty-year-old married couple. The quotidian violence that characterizes the couple's lives is provoked by fear. Piñera's characters, Tota and Tabo, represent a domestic dispute that mirrors the spectacality of the political context that was taking place in Cuba off-stage in the late 1960s. This violence is sparked by a fear that characterizes Tota and Tabo's existence and comes between them and the outside world. Their fear is triggered by the knowledge that everything about them—including their intimate secrets—is known by another. This comprehensive fear is embodied in a *planilla* the couple is forced to confront in the second act. The *planilla*, in *Dos viejos pánicos*, is a governmental survey that becomes the personification of fear for the middle-aged couple. Their violence against one another and against the outside world can be witnessed in their struggle with, and attempted elimination of, this very *planilla*. The internal world of Tota and Tabo reflects the national context (that of Cuba in the late 1960s, but also the larger Latin American political and social context) and portrays a growing fear and the violence that it provokes. The small, claustrophobic existence that the couple describes is doubled by the political events that surround the writing of *Dos viejos pánicos*, a context in which the spectacularity is mirrored by Piñera's work.

Outside Cuba, the political and economic situation of the late 1960s and early 1970s in Latin America was experiencing an instability that contributed to a time marked by change and unrest. The year before Piñera wrote *Dos viejos pánicos*, Che Guevara was killed in Bolivia while attempting to spark a revolutionary uprising along the models of the triumphant Cuban one. In 1968, the very year that this play was written, the Mexican President, Gustavo Díaz Ordaz, ordered the massacre of protestors, many of them students, in the Tlatelolco district of Mexico City. Both events highlight the unstable international environment in which Piñera was writing *Dos viejos pánicos* and underline his notions as to the centrality of fear in everyday life.

The beginning of the Cuban Revolution marked a time of tremendous creative production and growth that came as a welcome relief to the years of Fulgenico Batista's control. Batista's March 10, 1952 coup d'état ushered in a time marked by constant plots and rumors, all of which Batista answered with censorship and the suspension of many democratic rights. The Revolution of 1959, in turn, was defined by openness and possibility both within the theater and the arts, more generally. This can be seen in the return to Cuba of many exiled artists from the 1950s and the prominent positions that they and other artists began to occupy. They had left Cuba in the 1950s in response to Batista's restrictions on democracy and the opposition. While Virgilio Piñera had already returned to Havana from Buenos Aires in 1958, many other Cuban intellectuals like Heberto Padilla, José Triana, Antón Arrufat, Nicolás Guillén, and Alejo Carpentier rushed back to Havana in the wake of the Revolution's triumph. Many intellectuals saw the Revolution as the long-awaited fulfillment of the promises of independence from more than a half century earlier.

The decade of the 1960s was both a time of hope and anticipation as well as one of restrictions and limiting definitions. While the main goal of the government was to establish a stable revolutionary government and economy, all sectors of Cuban life were affected by the political and economic decisions on the island. On the one hand, writers and artists received much more attention and government support than they had before the Revolution. On the other, in exchange for this support, they were to defend the Revolution and its practices or suffer in silence and oblivion. This expected allegiance was often antithetical to the artists' way of thinking, as seen in the case of Padilla. In 1965, the Cuban Communist Party was formed, consolidating much conflict between old and new communists. From 1966 to 1970, the focus turned to the construction of both a social consciousness and economic development, underscoring the need to create a people and a solid nation that would be able to inherit the Cuban project that was being imagined.[1]

In the area of the arts, the early years of the decade were marked by a new emphasis on their importance, particularly theater given its accessibility to large amounts of people. Following the Revolution's triumph in 1959, the theater in Havana and in the other provinces enjoyed more freedom and support from the new government.[2] Its position as avant-garde and open to the people were two elements the new government wanted to use in order to bring the Revolution and its ideas to a larger public.[3] Given the Revolution's emphasis on the masses and their access to all governmental services, theater was initially seen as one of the most accessible of the arts for both the people and the Revolution's message to question the old ways.[4] In this way, theater reflected and benefited from the ideological changes that characterized Havana and Cuba in the early 1960s. Abelardo Estorino himself remarked upon these new opportunities in theater in an interview with me in May of 2007: "Todos estrenamos masivamente. Yo, por ejemplo, escribí *El robo del cochino* y se estrenó inmediatamente. Inmediatamente. Todas mis obras se han estrenado durante su tiempo por otros directores hasta que yo tuve un fracaso con mi primera obra de Milanés [We all premiered our works in great number. I, for example, wrote *El robo del cochino* and it premiered immediately. Immediately. All of my plays have premiered in their time by other directors until I had a failure with my first play on Milanés (*La dolorosa historia*).]"[5] This high number of staged works was not only an aspect of Estorino's theater but of many playwrights, whose work formed a part of the national stage, and also of those in the theater community. The number of theater performances increased drastically in the first years of the Revolution.[6] The immediacy of theater allowed it to respond quickly to the innovations taking place in the political and social contexts and experimental theater benefited most.[7] This boom in theater activity, the result of what was officially seen as theater's ability to reach out to the masses, was reflected initially in Havana by increased representations and productions. This was observed in both the professional arenas (as seen in the boom of plays written by Estorino, Piñera, José Brene, and Eugenio Hernández Espinosa, among others) and in amateur productions and sparked theatrical groups outside Havana in the second half of the 1960s, such as the Conjunto Dramático de Oriente in 1966, Teatro Escambray in 1969, and the Cabildo Teatral in 1970.

However, the conflict in the arts between creativity and allegiance began to be strained as the decade continued. The tensions between creating the political atmosphere and the revolutionary citizen of the new Cuba came together in 1968 with events that would shape the Cuban cultural context of the following years. Stress from the awarding in this year of the

Unión Nacional de Escritores y Actores de Cuba [National Union of Writers and Artists of Cuba (UNEAC)] prize to Antón Arrufat's play *Los siete contra Tebas* and Herberto Padilla's collection of poems' *Fuera del juego* culminated in what is known as the *caso Padilla*. This was discussed in detail in the introduction but bears a quick review here. Despite the, in the words of the UNEAC, 'counterrevolutionary' ideology of these two literary works, the two were prized by the UNEAC, though both authors suffered censorship and alienation from Havana's intellectual circles. Arrufat was relegated to working in a municipal library, exiled from the theater world, while Padilla was incarcerated in 1971 for his counterrevolutionary literature. This imprisonment sparked an international condemnation of the Cuban Revolution by such staunch supporters as Jean Paul Sartre, Octavio Paz and 80 other international writers and artists. Many of the social and political pressures from these events came to a head in the Congress on Education and Culture in this same year, where new, harsh regulations were enacted in order to control and monitor the university and artistic communities. Most importantly, publication standards were created to dictate the essential revolutionary quality of all works published or prized in Cuba. In this way, the Casa de las Américas, which held an influential place in all of Latin America, began to award works that were clearly political and revolutionary in nature, whereas in the 1960s many of the prize winners were not so openly political in nature.[8]

As seen in the above example, censorship in Cuban arts production permeated both official and unofficial efforts.[9] The literary communities in the decade of the 1960s both enjoyed a new freedom and security but also a need to define oneself in terms of the Revolution, all of which would affect the theater being written and produced.[10] This connection to new revolutionary definitions can be seen in Fidel Castro's words in his 1961 speech "Palabras a los intelectuales [Words to the Intellectuals]:" "dentro de la Revolución, todo; contra la Revolución, nada [within the Revolution, everything; against the Revolution, nothing]."[11] Indeed, Castro defended this need to curtail certain freedoms in pursuit of the Revolution. As Hugh Thomas quotes in his study of Cuba, Castro defended the need to define truth as an ingredient in the revolutionary cause: "These gentlemen who write 'truth never hurts', I don't know whether they conceive of truth as an abstract entity. Truth is a concrete entity in the service of a noble cause."[12]

Staging and censoring theater in the first few years following the Cuban Revolution was an issue that changed according to the goals of the official governing body. The same openness that would be seen as a revolutionary attribute of theater would also subject it to swifter censorship since all artistic genres became vulnerable to Castro's "Palabras a los intelectuales." Theater originally enjoyed an increase in access to space and performance.

However, after 1962 this decreased significantly and virtually ended by the end of the decade with events such as the Padilla Affair, as can be seen with the suppression of Arrufat's play *Los siete contra Tebas* and the resulting actions against the playwright. The early freedoms the theater community experienced after the Revolution came to an end as the decade of the 1960s closed, a fact that Rine Leal, a leading Cuban theater critic of the period, connects with the "the moralistic persecution of the artists" and "the malignant discrimination on the part of some cultural sectors against the artist in general."[13]

While there are many differences between the two countries and times, one can ask how much Cuban censorship shared with Soviet practices of the 1920s or the 1930s.[14] This was a connection that was often discussed by those experiencing firsthand these processes of censorship. Guillermo Cabrera Infante, an early supporter of the Revolution who then broke with it, makes a parallel with Stalinism in his memoirs *Mea Cuba* (1992).[15] Here he recalled an event from a 1961 meeting between Cuban intellectuals and Fidel Castro where Virgilio Piñera made a confession of fear. In the words of Cabrera Infante, he said: "'I think it has to do with all this.' It seemed that he included the Revolution in his fear, though apparently he meant only the crowd of so many so-called intellectuals. But perhaps he was alluding to the life of a writer in a Communist country: a fear called Stalin, a fear called Castro."[16]

Just as the Soviet Union under Stalin increased control over cultural production, Cuban control mechanisms throughout the 1960s and into the 1970s emphasized more and more how the arts were to benefit the people. What did not fit into this definition was considered against it as can be seen in Castro's "Palabras a los intelectuales." Nevertheless, despite the commonalities between the countries, Cuba and its processes of control were not just a copy of the Soviet model. In fact, Cuba strove to differentiate itself from the Soviet and U.S. imperialism in the wake of the Cuban Missile Crisis of 1962. Though Cuba always tried to maintain its independence, this difference was de-emphasized after the failed *Zafra de los diez millones*[17] of 1970 when Soviet aid was needed.[18]

On the topic of control and censorship in Cuba, we need to remember that virtually every apparatus of cultural production was controlled by the state. Publishing houses came under the control of the Institute of the Book in 1967. This was the same year that copyright was abolished, a move that made works part of the state's property. All manuscripts were read by government functionaries and all artists were paid by the state. While this freed the artists from market fluctuations, it also increased their dependence on what the state deemed publishable.[19] These changes can be seen to help the arts community in Cuba, but they also increased the state's intervention and control over artistic production, making it difficult for some to publish. This difficulty can be seen in both the lives and work of writers such as Reinaldo Arenas and Piñera, himself. Due in part to the fact that he was gay, Arenas had great difficulties publishing in Cuba in the 1960s and 1970s and had to have his *El mundo alucinante* (1968) smuggled to France to be published. Piñera was arrested at his home in 1961 during a purge of men suspected of homosexual behavior and spent the night in jail. Thanks to his friends he was released quickly, but the arrest was said to have affected him deeply.[20]

Another method of control that also increased the visibility of Cuban artists was the new literary prizes that had been established after the Revolution, such as those awarded by the UNEAC or Casa de las Américas.[21] This control can be seen in the reaction to the works of Heberto Padilla's *Fuera del juego* and Antón Arrufat's *Los siete contra Tebas*. A work that was not seen as "revolutionary" could win the prize but be published with a letter underlining its counterrevolutionary status or immediately taken off the shelves. A move like this ostensibly emphasized the openness of the Revolution to criticism while not allowing the public to access that criticism.

For well-known members of the intellectual community who could not justifiably be silenced, there was another option that would allow their works to be published but in limited numbers. This can be seen with José Lezama Lima's masterpiece *Paradiso* (1966).[22] Only four thousand copies of this neobaroque novel were published, even though Lezama Lima was both internationally and nationally known as a central figure of the literary tradition in the twentieth century. The only plausible reason for this is what officials saw to be amoral and counterrevolutionary in the novel's depiction of gay sex.

Virgilio Piñera's *Dos viejos pánicos* was written within this political context full of tension and change, when artistic creations were often judged on their perceived political commitment and was both the beneficiary and the victim of many of these events. The play won the Casa de las Américas prize for theater in 1968, nine years after the triumphant entrance of Fidel Castro in Havana, but did not premiere in Cuba until 1990.[23] The play is divided into two acts and portrays a day in the life of Tota and Tabo, each described as "vieja [old woman]" and "viejo [old man]" around sixty years old.[24] Fear becomes an omnipresent element that invades and, in turn, guides the direction of their lives and of the play. Because of this fear, violence begins to emerge between the two characters, as it is through and by means of violence that they relate to each other. Virgilio Piñera gives central stage to fear and violence in order to question the role that they both occupy in Cuban society at the time. In *Dos viejos pánicos*, Virgilio

Piñera portrays fear as *the* element that permeates and corrodes our relationships to one another. This fear produces a violence that punctuates our actions and desires, and can only be stopped through its violent repression, creating an interminable circle of oppression that the playwright wishes to end.

Virgilio Piñera is a central literary figure of the Cuban canon of the twentieth century. Unlike the other three playwrights studied here, Piñera is as well known for his narrative, poetry, and critical work as for his dramatic work. Initially, he entered into the literary community through critical contributions to journals such as *Orígenes* and *Espuela de Plata* both published in Havana, and through his poetry collections, *La isla en peso* (1943) and *Vida de Flora* (1944). The themes present in these works hint at the irregular way Piñera viewed the Cuban society and canon and the discomfort that Piñera felt being aligned with a literary group or tendency, topics that would remain central in his work.

Piñera left Havana for Buenos Aires in 1946, where he stayed until 1958, only returning to Cuba briefly during these years. This time was marked by his involvement in the literary community of the Argentine capital and a friendship with Witold Gombrowicz. He collaborated on the translation of Gombrowicz's first novel *Ferdydurke*. In Buenos Aires Piñera also contributed to literary magazines and associated with members of the influential *Sur* group, such as Victoria Ocampo and Jorge Luis Borges. His narrative work, much of which was written in the decade of the 1950s is comprised of both short stories, *Cuentos fríos* (1956) and novels. His first, and without a doubt, best known novel is *La carne de René* (1952) and considered by many to be his masterpiece. Two of the central themes that have been identified with Virgilio Piñera's narrative and poetry are insularity and corporality, topics that can be seen in *La isla en peso* and *La carne de René*.[25]

Though he continued to write throughout his life, Virgilio Piñera was largely unpublished during the last years of his life and beyond, because of the strict censorship and control of the Cuban literary apparatus during the 1970s. Literary criticism of his narrative, poetry and theater started to revive in the late 1990s and early 2000s, particularly after the 1992 symposium in honor of Piñera that took place in Havana. At the same time, the publication in Cuba of a number of important works, such as Antonio José Ponte's *La lengua de Virgilio* (1992), refocused Piñera in the present moment while others like Antón Arrufat's *Virgilio Piñera: entre él y yo* (1994), gave an account of his conflictive last years.[26] More recently, his status as an important figure in the Cuban literary canon has been even more highlighted with the 2001 issue of the journal *Encuentro de la cultura cubana* dedicated to Virgilio Piñera, and with Rita Molinero's 2002 edited collection of various essays on Piñera and his literary work. Enrico Mario Santí in "El fantasma de Virgilio," part of *Bienes del siglo: Sobre cultura cubana*, details the life of Virgilio Piñera with an eye to understanding his literary works. This essay also comments on the last years of his life spent in silence thanks to the *quinquenio gris* and the aftermath of the *caso Padilla*. Thomas F. Anderson has combined an exploration of both Piñera's life and his literary production in his recent *Everything in Its Place: The Life and Works of Virgilio Piñera* (2006). This pivotal book examines his life in connection with his many, varied literary works and is an invaluable contribution to studies on Piñera and twentieth century Cuba.

However, this recent canonization of Piñera is not without its conflictive side, as Piñera has oftentimes been read as a figure who unsettles a Cuban literary canon that includes such names as José Lezama Lima, among many others. In fact, Piñera himself discusses the idea of a national literature in his own essays on Cuban literary production both before and after the Revolution of 1959. Ana García Chichecter, in the article "Virgilio Piñera and the Formulation of a National Literature," analyzes Piñera's writings on Cuban literary tradition expressed both in poetry and his essayistic investigations on the same topic. She takes as a point of departure his talk "Cuba y su literatura" delivered at the Havana Lyceum on February 27, 1955 and published later that year in the journal *Ciclón*, an important journal edited at the time by José Lezama Lima. In García Chichester's interpretation of Piñera's words, she sees that he criticizes "the absence of critical inquiry based on the *ethical reading of texts*."[27] This was not a criticism of "the insufficient talent on the part of Cuban writers but results from a deficiency of another kind: the need to define themselves in terms of their difference."[28] Piñera wanted Cuban literature to emphasize what made it different, what defined it in light of other literatures. García Chichester identifies how he did this with his own work, especially the poems "La isla en peso" and "Vida de Flora." These poems, like all of Piñera's work, show a desire to move beyond the facile definitions of the topics he portrays and to delve deeply into their meaning. Similarly, he urged the Cuban literary community to do the same in its work, both creative and critical. This was an aspect of the Revolution of 1959 that he originally championed, though his views changed as those of the official government did. In this way, Virgilio Piñera was not a figure comfortable with what he saw as the traditional literary canon of Cuba and, for this reason, he can be seen as an uneasy addition to the canon of Cuba's twentieth century.

Virgilio Piñera established himself as an important Cuban playwright of the twentieth-century with his early *Electra Garrigó* (1941), which melded Greek myth with Cuban cultural topics, as seen in the very title. His contribution to Cuban theater continued throughout the following decades until his death in 1979. Many of his plays are identified within the definitions of the Theater of the Absurd (a definition that will be explored later on in this chapter), with the exception of *Aire frío* (1959). This play is a more realist portrayal of a Cuban family living in reduced circumstances from the years 1940–1958. The title refers to the suffocating heat that marks the work. Much of Piñera's other theater, though, is not tied directly to Cuba, though there may be allusions to the island in the text. Instead, Piñera focuses on portraying the frustrations and indignities that characterize the modern existence. These can be seen in works such as *El no* (1965), *La niñita querida* (1966), and *La caja de zapatos vacía* (1968).

Dos viejos pánicos is arguably his best-known play of the 1960s, given the many issues it touches upon and the recognition from the Casa de las Américas prize. It is considered by many to be the play that officially recognized Piñera's centrality in the Cuban and Latin American theater community. As quoted in the Introduction to this study, Rine Leal stated that *Dos viejos pánicos* was the work that garnered official acknowledgment for Piñera's extensive influence on Cuban theater, an assertion with which Raquel Carrió Mendía agrees.[29] For her, *Dos viejos pánicos* is the culmination of both a theater of the vanguard and of Piñera's theater itself.[30] This play is seen as the pinnacle of Piñera's dramatic work—work that spans three decades—and a central moment in Cuban, and Latin American, theater that demands an in-depth exploration of its contribution. Given the importance that these two critics give to Virgilio Piñera's dramatic work and *Dos viejos pánicos* for Cuban theater, this play is a logical foundation in order to appreciate the location of violence in Latin American theater in the 1960s and to explore the directions in which it moved in the 1970s and beyond.

Dos viejos pánicos, like much of Piñera's theater, has traditionally been classified as Theater of the Absurd, a theater movement that originated in Paris in the 1940s and 1950s. The Theater of the Absurd was born from a desire to reform theater in the wake of the destruction of World War II and its aftermath.[31] The notion of the absurd as a way to understand the post-war situation of humankind originates out of Albert Camus and Jean Paul Sartre, who believed that Man should accept the ultimate lack of meaning of the world. As discussed quickly in the Preface, Antonin Artaud and Martin Esslin, two of the movement's central theorists, theorized on what they viewed as the absurdity of the world. Artaud, known for his Theater of Cruelty which shared much with Theater of the Absurd, and Esslin, viewed theater as the means to provoke its audience beyond traditional ideas and to, thus, change what would happen both in the theater and outside it.[32] According to these thoughts, theater was not a purely entertaining art form but was meant to awaken the spectator to what was taking place in the world. This was a move way from Aristotelian and Renaissance theater and blankly pointed out the absurdity of the modern world. Eugene Ionesco's *The Bald Soprano* (1950) and Samuel Beckett's *Waiting for Godot* (1953) are classic examples of how the Absurd portrayed the anxiety towards life and death and the ultimate irrationality that these writers believed characterized the world.

The ideas that the playwrights and theorists of the Theater of the Absurd in Europe put forth were quickly adapted by many Latin American dramatists, though this has also been a point of contention in Latin America. Theatre of the Absurd is a term that has traditionally been used to describe this genre in Europe while some theorists use the term Absurdist theater to differentiate the phenomenon in Latin America. Absurdist theater has been seen as such a variation on the European model by some critics that it deserves different terminology. Daniel Zalacaín and Raquel Aguilú de Murphy see Absurdist theater as focusing on the same human alienation through violence and the manipulation of language, though it is in the political-social topics that the absurdist playwright differs from the European.[33] Other critics, such as George Woodyard, Terry Palls, and Eleanor Jean Martin, have not seen a need to create such a difference in name in the two manifestations and use the term Theater of the Absurd for this school of thought in Spanish America.[34] Woodyard, in his 1969 essay "The Theatre of the Absurd in Spanish America," examines how this European Absurd translated to the theater of Spanish America in the 1960s. There was of course a change in the geographical context, though the focus on irrationality and fragmentation remains central. Woodyard identified the most important elements of the Spanish American Theater of the Absurd as the following: plays with two characters, anti-heroes, physical violence stemming from feelings of contempt and hatred, and an insistence on fragmentation.[35]

Terry L. Palls and Eleanor Jean Martin both turn their attention specifically to the role of Theater of the Absurd in the context of Revolutionary Cuba, defending the Absurd against contemporary accusations that saw it as essentially anti-revolutionary. Palls identifies the Absurd as a minor but fundamentally important influence on Cuban Theater in the first decade of the Revolution.[36] The central difference between Absurd dramatists and those traditionally seen as politically committed, he explains, is the emphasis on technique as well as the artistic quality of the theater, as opposed to an emphasis on content.[37] Thus, it can be argued that the dramatists that were more influenced by the Absurd were the ones who were innovating and renewing Cuban theater, despite their apparent lack of Revolutionary commitment. In fact, Eleanor Jean Martin argues for the basic revolutionary quality of the Theater of the Absurd when employed by many Cuban playwrights: "the theater may be thought of as positive, for it helps man to understand fully what is happening to him, in a transitional society. In the tradition of the Theater of Cruelty, the spectator feels that he is seeing the essence of his own being before him, that his own life is unfolding within the bodies of others."[38] For Martin, rather than being anti-revolutionary, the Theater of the Absurd, and specifically Piñera's employment of it in *Dos viejos pánicos*, shows the concrete reality of transition that the Cuban society is undergoing in order to become a revolutionary one. Theater becomes the space where the public can experience changes through the actions of the characters.[39] Theater of the Absurd, then, can be viewed as the ultimate revolutionary space for innovation and progress.

Diana Taylor, in turn, uses the term theatre of crisis to describe the theater that she examines from the years 1965 to 1970 in her book *Theatre of Crisis: Drama and Politics in Latin America*. Theatre of crisis differs from the absurd or protest theatre, in her words, because in this theater "both the objective context and the subjective consciousness threaten to collapse."[40] The absurd, in contrast, "reflects a crisis ideology, the personal sensation of decomposition, within a stable, bourgeois context."[41] For Taylor, there is an insistence on the part of North American scholars to overemphasize the use of European terms within the Latin America, a context that does not always lend itself easily to European definitions. This is an important point that needs to be remembered in reference to Latin America. These two regions are different, calling upon distinct historical and social traditions that are not always in correspondence. Furthermore, it would be unwise to believe that the European traditions alone have had an influence in the creation of Latin American theater. As was explored in the preface, the Cuban tradition of *teatro bufo* is an essential element in the Cuban absurd and in Virgilio Piñera's theater. Nevertheless, it is necessary to remember that European theater was read and performed in the cities of Latin America and one can see a connection between the themes of the theater of the two regions. This is not to say Latin American absurd is an exact copy of the European or that the topics of Beckett's and Ionesco's theater will be seen unchanged in Piñera's or Jorge Díaz's work, for example. These playwrights innovated and adapted the absurd to their own contexts in order to create a Theater of the Absurd that fit their present circumstances and contexts, as we can see in the example of Piñera's theater, particularly *Dos viejos pánicos*.

Virgilio Piñera's Theater of the Absurd as manifested in *Dos viejos pánicos* is, in fact, the site of questioning from which the spectator can understand the conflicting pressures that manifest in the modern being, given his focus on the universality of the manifestation of fear and violence. As Piñera himself states his theater is not "del todo existencialista ni del todo absurdo [neither completely existentialist nor absurd]."[42] Piñera's early examples of Theater of the Absurd, such as *Electra Garrigó* and *Falsa alarma* (1948), predates the pivotal example of the European Absurd, Ionesco's *The Bald Soprano*. Nevertheless, Piñera wrote from the period in which he lived and was influenced by the same events as the European absurd writers. These differences in context and years do not change the similarities between the European and the Latin American absurd nor does it conflate them into one movement.

Dos viejos pánicos is a play that is symmetrically, almost obsessively, divided into two acts with two characters, who reflect each other in turn by the alliterative repetition of their names, Tota and Tabo. Piñera's use of these two names is already an indication of the reflective relationship that they have with one another.[43] In the first act, these two protagonists begin a sinister game that they repeat daily: "killing" each other (metaphorically), and describing the fearless existence that they imagine their deaths would provide. Tabo, in an indication of the violence to come, violently cuts out figures of young people from a magazine in order to later burn them, while Tota attempts to convince him to play her game of dying, by forcing him to confront his own aged reflection in a hand mirror—an action that will force him to see his progress toward old age and death. They bicker about the merits of their respective pastimes but, using the mirror and his own self-reflection as a threat (and Tabo does view it as a threat), Tota persuades Tabo to "strangle" her. After he does, the two enact how their lives would be after her death. Tota describes her life free of fear and consequences: "Cuando uno está muerto ya no hay consecuencias. La última fue morirse [When one is dead, there are no more consequences. The last one was dying]" (Act 1: 484). The two characters act out what life would be like after Tota's death, describing their imagined lives free of fear and the limitations it imposes. The couple's verbal exchanges in this act are short and to the point, demonstrating the extent to which they know one another's weaknesses and the best way to exploit them. Because they know one another so well, their conversation is full of allusions to their long history together, making difficult a quick and easy understanding of their dialogue and the play's direction. Despite the complexity in Tota and Tabo's conversations, it is clear that they have tired of each another and use their mutual familiarity to provoke one another.

The second act opens over a daily glass of milk, where Tota and Tabo discuss the quotidian aspects of their life, including their "death" game and Fear. At one moment, they remember the *planilla* that they are to fill out and Tabo begins to ask Tota her questions.[44] The activity only inspires more fear, given the subject matter and the fact that the *planillero* knows all about them. They begin to fight over the *planilla* and, in their struggle, "kill" one another. It is in this state that they decide to kill Fear, an act that ultimately proves to be impossible and they return in the end to their everyday existence where tomorrow, as Tabo says, is "[o]tro día más [another day]" (Act 2: 509). This focus on routine emphasizes the circularity that defines their lives despite the extremity of their earlier acts.

The conversations and actions of the play, though difficult to understand at first, demonstrate the repetitiveness present in the couple's life and the fine-tuning that they have done in this one day that they repeat again and again. There is much that is left unsaid in Tabo and Tota's dialogue and actions, thus underlining the amount of time that they have dedicated to fine-tuning this process. *Dos viejos pánicos* is a play that leaves as much unsaid as that which is actually said. The couple communicates through gestures and attacks, having perfected their nonverbal communication over the years. In this way, multiple readings reveal the subtleties that characterize their relationship and their fears. From this fear of life and all that it entails comes a violence that defines their relationship to one another and their relationship to the outside world. In order to stop the insecurity that comes from their fear, they turn destructively violent against one another, and the two of them together turn against youth, against authority and, in vain, against fear.

Violence in *Dos viejos pánicos* tries to hide the profound fear that the two characters experience. The title, *Dos viejos pánicos*, is of particular importance in the play and to this study given the many meanings it can have. For Matías Montes Huidobro one of the defining elements of Cuban theater is the employment of schizophrenic elements. For Virgilio Piñera and the absurd, the use of schizophrenic language and the play on words that can be seen in the characters' names and in the very title sets this play apart.[45] As José Corrales highlights, it is not clear whether it is referring to two old frightened people or two old fears, thus creating doubt as to what is being described or what is doing the describing: "Entonces, ¿se refiere el título a dos personas viejas, cada una de las cuales es 'un pánico'? ¿O hay dos pánicos (miedos) que han durado muchos años (son viejos)? ¿Está el adjetivo sustantivado o es el sustantivo el que se una como adjetivo? [So, does the title refer to two old people, each one of whom is *'un pánico'*? Or are there two *pánicos* (fears) that have been around for many years (are old)? Is the adjective made into a noun or is the noun that which becomes an adjective?]"[46] Adjectives that turn into nouns, or nouns that turn into adjectives, play on the doubling mechanism of the play in a grammatical sense. That is, the grammar in *Dos viejos pánicos* seems to play on the doubling that is seen in the couple itself. As Merlin H. Forster maintains, the emphasis is placed on the two human figures if the noun is *viejos* and on the fear if the noun is *pánicos*.[47] Thus, it is virtually impossible to determine without a doubt whether the center of the play will be two human figures or the more nebulous and ancient Fear that becomes almost a third character. The ambiguity that can be identified in the very title underlines the elusiveness of language in the play in general, and hints at the multiplicity of readings that this play can accept and that have been put forward of it.

The Dictionary of the Real Academia Española defines "pánico" first as "referente al dios Pan [referent to the god Pan]." The second definition found in the RAE dictionary is "del miedo extremado o del terror producido por la amenaza de un peligro inminente, y que con frecuencia es colectivo y contagioso [the extreme fear or terror produced by the threat of imminent danger, which is frequently collective and contagious]."[48] *Pánico*, then, describes an immense fear that is often infectious to a larger collective of people. Given that the play consists of just two characters, this suggests that it is referring to the existence of an outside community from which the couple may have been infected by, or which the couple may infect with, fear. This last element—that of the infectious nature of *pánico*—is significant, given that it suggests an inability on the part of the infected to control their *pánico*. Instead, they are the victims of an outside force that overpowers them. Furthermore, this terror is the product of an "amenaza de un peligro inminente [threat of imminent danger]," identifying the threat as something that has not come to pass but hovers over the panicked with just the promise that something may happen. This definition is important to the play's analysis since it offers insight into the characters and their construction.

By placing *pánico* in the title, Piñera hints at the presence of Fernando Arrabal's Panic Theater from the 1960s, an allusion which both Corrales and Merlin Forster reference in their essays.[49] Arrabal, together with Alejandro Jodorowsky of Chile and Roland Topor of France, began to formulate many of the ideas that later formed the Panic Movement, and then Panic Theater, in 1960 in Paris. In February 1962, the group decided to use the word "*Panique* (Panic)" to define their movement rather than the earlier "Burlesque." With this new term, the group hoped to continue encompassing the ideas of the baroque, the contradictory, the mythical, the monstrous, while simultaneously embracing many new connotations from the word 'panic' itself. Primarily, panic refers to a collective or individual feeling of overwhelming terror that is oftentimes irrational. Additionally, as seen in the definition quoted earlier, the word "panic" is etymologically linked to the god Pan, who was the Greek god of fields, forests, wild animals, flocks, and shepherds. Physically, he resembled a goat and has been associated historically with sexual orgies and debauchery. This feeling of panic, of fear, was inspired by Pan. Arrabal himself made use of a definition put forth by Joseph Campbell in order to explain the Panics' appropriation of Pan and the word panic:

> La emoción que provocaba en los seres humanos que por un accidente se aventuraban en sus dominios era el terror "pánico," un terror repentino y sin causa. *Cualquier motivo trivial, una rama que se rompe, el movimiento de una hoja, hará que la mente se estremezca con un peligro imaginario y, en el esfuerzo enloquecido para escapar de su propio inconsciente despierto, la víctima expira en su fuga aterrorizada.*
>
> [The emotion it provoked in human beings that by accident adventured into its domain was one of "panic" terror, a sudden terror without cause. *Whatever trivial motive, a broken branch, a leaf rustling, will make the mind tremble with an imaginary danger and in the crazed effort to escape from his own awakened unconscious, the victim will expire in his terrified flight.*][50]

This almost hysterical feeling of terror that in the ancient world came from a brush with Pan is the emotion that Arrabal and the other Panics hoped to inspire in their public. In this way, many of their events, though initially formed from a script, were improvised, giving the representations a transient quality.

For Arrabal and Jodorowsky, the objective of Panic Theater was to provoke the spectator to action, not to create an enduring work of art.[51] Panic Theater shared much of the same ideas that had earlier inspired the Dadaists, such as the idea that the Panic should propel one to euphoria and collective celebration. Although these ideas also overlap with those of the Surrealists, the Panic movement, whose members had been close to Andre Breton in the early 1960s, was formed as a break from the dogmatisms that they identified in Breton's surrealism in the 1960s.[52] It was conceived as a tolerant, eclectic, and constructive movement that advocated liberty of thought and action.

Virgilio Piñera's use of the word "panic" and its allusion to Arrabal and Panic Theater in *Dos viejos pánicos* is an interesting reference that helps to reveal much about the play and offers insight into its direction. Primarily, the fact that Panic Theater was closely tied to theater over other art forms is an important detail that underlines Piñera's use of *pánico*. Furthermore, one of the fundamental aspects of Arrabal's Panic Theater is its origin in a freedom from censorship and any restricting elements. This is an important connection both within the argument of the play and the context which surrounded Cuban intellectual culture in the late sixties. For both Tota and Tabo, the panic that the outside world provokes in them is exactly that which restricts their actions and censors their ability to act. For Arrabal, Panic Theater creates the unexpected by using memory and chance.[53] That is, the combination of the memory of historical and biographical events combines with the confusion of unforeseen acts. When the unexpected becomes what is most feared, the spectator can recognize a clear opposition between the goal of the play and the life of the characters within it. This provocation is exactly Piñera's goal in using the word *pánico* in his play. That is, *Dos viejos pánicos* wants to incite the spectator to action, to thought through a presentation of Tabo and Tota's fears. *Pánico*, as in an intense fear, refers to the emotion that both protagonists experience in their everyday lives. Its meaning goes beyond fear, to recall Arrabal's use of the word as a "terror repentino y sin causa [sudden terror without cause]" provoked by the actions of the god, Pan. Panic Theater and Piñera in *Dos viejos pánicos* use this intense emotion to incite a celebration that will force the characters and spectators to move beyond their fear and inability to act. Instead, they will be inspired to question their surroundings and their own actions—the theater will wake them up to an existence beyond fear and give them the ability to live in a state of engagement with their environment.

When considering the use of the word *pánico* in the title and the focus on fear throughout the play, it is significant to remember the incident reported by Guillermo Cabrera Infante about Virgilio Piñera in the early days of the Revolution that was referenced earlier in the discussion of censorship. It is important to return to this in light of the question of fear. According to Cabrera Infante's *Mea Cuba*, in 1961 during a meeting of prominent Cuban writers from *Lunes de Revolución* and *Revolución* with Fidel Castro and members of his government, Piñera confessed to being afraid. I quote the moment at further length here to understand Piñera's own fear:

> Suddenly, out of the blue alert: a timid man with mousy hair, frightened voice and shy manners, slightly suspect because he looked frankly queer in spite of his efforts to appear manly, said that he wanted to speak. It was Virgilio Piñera. He confessed to being terribly frightened. He didn't know of what, but he was really frightened, almost on the verge of panic. Then he added: 'I think it has to do with all this.' It seemed that he included the Revolution in his fear, though apparently he meant only the crowd of many so-called intellectuals.[54]

As explored earlier, this is a significant anecdote about the relations between Castro and the Cuban intellectuals that would worsen considerably at the end of the 1960s and beyond. However, it also provides an interesting insight into Piñera's frame of mind in the decade following the Revolution and allows us to speculate on the later use of fear in his award-winning play. By publicly admitting to his own fear in front of government officials and fellow intellectuals, Piñera gives center stage to fear in Cuban society. He uncovers something that is often not admitted publicly but buried under an outward need for courage and bravery. In both the above anecdote and the play *Dos viejos pánicos*, Virgilio Piñera makes us reflect upon the far-reaching effects and consequences of an emotion that many prefer to ignore while forcing the spectator to consider the misdeeds realized in an effort to conceal fear and the other many actions it impedes. Admitting to fear, then, is the first step to being set free from it and to being able to act from a space beyond fear.

It is precisely fear and the provocation that Piñera identifies with it that underlines the violence in *Dos viejos pánicos*. Violence here does not take the shape of war in the sense of a national crisis (as we see with Estorino or Gambaro). Instead, the violent acts between and around Tota and Tabo come from a more domestic, psychological source. The violence between the two characters takes place through their words and their frustrations with one another. In this way, the characters become a sort of double of both each other and, more importantly, of the spectators. The audience sees characters onstage that offer an opportunity to consider how violence defines one's relationships to those closest to one and to the outside world. Violence here becomes a mechanism that Tota and Tabo use to confront and react to each other and to the outside world. Piñera creates these characters not so that the spectators will identify with them but, instead, so that the audience will be forced to ponder their own quotidian use of violence in their personal relationships. This observation departs from many other considerations of violence given that we are not witnessing the acts of a war or a political struggle, but the everyday use of violence in our lives.

Dos viejos pánicos opens with both characters on stage; Tabo has his back to the audience and Tota is facing the audience. Tabo is engaged in what seems to be his favorite pastime: "recortando figuras de una revista [cutting out figures from a magazine]" (Act 1: 479). With this action, Tabo initiates his ultimate project, one whose very utterance has a performative quality that punctuates his desire and his words: "Recortar y quemar. Sí, Tota, hay que quemar a la gente. Ayer quemé doscientas, y hoy pienso quemar quinientas [Cut out and burn. Yes, Tota, I have to burn people. Yesterday I burned two hundred, and today I mean to burn five hundred]" (Act 1: 480). Tabo, who fears old age and hates youth, believes that through this action he can control the world's youth and his own aging process. To Tota's observation that the previous year 100 million children were born, Tabo responds,

> Cien millones…(*Pausa.*) ¿Y qué…? Los voy a quemar uno por uno. (*Pausa.*) Oye, Tota, óyelo bien: desde hoy, desde este momento, tendré mucho trabajo. (*Ríe.*) Muchísimo trabajo. (*Pausa. Cuando Tabo dice*

"*trabajo*," *Tota baja de la cama y, con el espejo oculto en la espalda llega junto a él*.) Quemaré diariamente cien mil recién nacidos, cincuenta mil niños de cinco años, treinta mil de diez, veinticinco mil de veinte y diez mil de veinticinco… [One hundred million… (*Pause*.) So what…? I'm going to burn them one by one. (*Pause*.) Listen, Tota, listen up: from now on, from this moment, I'll have a lot of work. (*He laughs*.) A lot of work. (*Pause. When Tabo says "work," Tota gets down from the bed and, with the mirror hidden behind her back, comes up to him*.) I'll daily burn one hundred thousand newborns, fifty thousand five-year-old children, thirty thousand ten-year-olds, twenty-five thousand twenty-year-olds and ten thousand twenty-five-year-olds] (Act 1: 482).

Inherent in this act, the reader can observe a strong level of violence and disregard for the other that underlies all of the couple's relationships. Simultaneously, this highlights their powerlessness against the world and the cowardice that can be seen in burning magazine figures. Tabo's act indicates a hallucinatory quality inherent within it given that what he is proposing to do is impossible. It is a performative gesture that in its utterance underscores desire as well as its impossibility. Ultimately, burning figures cut from magazines suggests a perverted violence—while it can be seen as a violent act, it also implies a disconnect with reality and an inability to face that which they most fear. Simultaneously, Tabo's actions point to the symbolic power that resides in the gesture of cutting up and burning figures that underlines the reality of the situation. Tabo tries to control the uncontrollable—both his own fear as well as the external world—and ends up highlighting his *loss* of reality.

Perhaps the most important manifestation of violence in this play appears in the relationship between the two characters. The text reveals a romantic history between the two that has, like any other relationship, weathered difficulties, and perhaps started to sour. Both Tota and Tabo, like many couples, show conflicting desires for the relationship and often turn to violence to solicit a wanted response from the other. For example, Tota wants Tabo to engage in her game and resorts to perverse tactics in order to convince him. She first threatens physical violence instead of helping him with the magazine photos: "¿Ayudarte a ti, precisamente a ti? La única ayuda que te daría sería un empujón. [Help you, you exactly? The only help I would give you would be a push!]" (Act 1: 479). Tota intensifies the conflict after Tabo taunts her with her past loves and failures, when, aware of Tabo's fear of old age, she tries to force him to look at himself in a hand mirror:

> *Tota*: (Con la mano izquierda le coge las manos a Tabo y las aparta de su cara al mismo tiempo que le pone el espejo frente a la cara.) Mírate.

Tabo: (*Se abraza a Tota tratando de quitarle el espejo, que ésta ha vuelto a ocultar en su espalda.*) Puta vieja, te voy a estrangular, qué te has creído. ¡Egoísta! Conque jugar al juego y Tabo que reviente, ¿no? Acabaste con mi paciencia, puta mala. Hoy es tu último día. (*Ruedan por el suelo y se detienen en el centro del círculo.*)
Tota: (*Que ha caído encima de Tabo, dejando el espejo sobre la cama.*) Y tú, viejo chulo, quemando jóvenes en efigie. ¿Por qué no los quemas de verdad? Anda, sal a la calle y empieza a quemar gente. (*Pausa.*) Eres un tipejo, Tabo, eso es lo que eres.
[*Tota*: (*With her left hand she grabs Tabo's hands and takes them away from his face as she places the mirror in front of his face.*) Look at yourself.
Tabo: (*He hugs Tota, trying to take away the mirror that she's hidden again behind her back.*) You old whore, I'm going to strangle you, what do you think. Selfish! So, you want to play a game and Tabo will break down, right? I've run out of patience, you stupid whore. Today is your last day. (*They roll around on the floor and stop in the center of the circle.*)
Tota: (*Who has fallen on top of Tabo, leaving the mirror on the bed.*) And you, you old villain, burning young people in effigy. Why don't you really burn them? Come on, go out and start to burn people. (*Pause.*) You're a piece of work, Tabo, that's what you are] (Act 1: 483).

The psychological violence that can be seen in this quote and the physical violence that it promises suggest a deterioration in the relationship and the humanity of the two characters that hints at the central project of *Dos viejos pánicos*. Here, Piñera indicates a universal situation of humanity where we harm those we love most precisely because we know their weaknesses and can exploit them. Thus, the role of violence in these intimate relationships takes center stage and suggests a desire on the part of the playwright to question this familial violence. Piñera is inciting the audience to consider how the couple's fear translates into violent acts that begin to erode and, then, define them, making the violent gestures between them the glue that maintains the relationship. Piñera wants to engage his spectators to re-evaluate their own use of violence in order to, then, put an end to the cycle of violent oppression that defines this relationship and many others.

Severino João Albuquerque's study *Violent Acts: A Study of Contemporary Latin American Theatre* underlines the possibility of ending violence by presenting a dramatic example that would serve as a deterrant, an idea that Antonin Artaud had proposed in his theater of cruelty. For Artaud's Theater of Cruelty, violence on stage would prevent any type of violence offstage, given that the horrific vision would hinder any chance that the viewer would want to replicate such a scene:

I defy any spectator [...] who will have seen the extraordinary and essential movements of his thought illuminated in extraordinary deeds—the violence and blood having been placed at the service of the violence of the thought—I defy that spectator to give himself up, once outside the theater, to ideas of war, riot, and blatant murder.["]55

In reference to Latin American theater after the Cuban Revolution, Albuquerque revisits Artaud's earlier assertion in his own study in order to understand the contemporary use of violence in this new community. In Albuquerque's opinion, violence onstage serves as a way to reflect on the human condition.[56] Thus, for both Artaud and Albuquerque, dramatic violence used onstage becomes a fertile space of dialogue used to counteract an external, social violence. Fernando Arrabal (whose work has already been discussed above as influential to *Dos viejos pánicos*) similarly sees a connection between violence onstage and debates offstage. For Arrabal, violence appears as an essential element in Panic Theater, given its demands for a spectacle that provokes terror in its public. Nevertheless, this terror is seen as constructive and fruitful given that from it the play will stir the spectator to liberation and action. Ultimately, Panic Theater is a theater of hope and promise: "El pánico es solo un *teatro de la esperanza*: una *esperanza lejana todavía* (frente a la *inmediatez de la esperanza revolucionaria*). [Panic Theater is alone *a theater of hope:* a *still distant hope* (opposite the *immediacy of the revolutionary hope*)."[57] It is my assertion that Piñera's objective is similar to the ideas proposed by Artaud, Albuquerque and Arrabal: through the couple's relationship, the audience can reflect on the role of fear and violence in their own life and society. The play causes the spectators to question and consider the circumstances before them. The violence is not meant to provoke more violence but instead to force a conversation on what is the place of violence in our lives and relationships. Thus, Piñera's theater becomes an instrument that intends to provoke a reaction and an awakening in its audience through its appropriation of fear towards life and outside authorities and its employment of violence between the characters.

Another important aspect that can be found in *Dos viejos pánicos* and relates to the practice of violence in the play is the circularity and repetition in Tota and Tabo's relationship with one another and the games that they play. Just as in José Triana's *La noche de los asesinos* (1965) where the three siblings continually rehearse their parents' murder, self-made games are repeated as a means to cope with the contextual surroundings. Primarily, as has been seen in earlier quotes, Tabo cuts out human figures from magazines in order to burn them, an activity whose frequency needs to be increased the more he performs it. Similarly, Tabo and Tota repeat regularly the same "killing" game, a part of their daily routine just like their evening glass of milk. This duplication suggests the couple's need to participate in an action that, outside the parameters of the game, they cannot reproduce. Their role-playing allows them to take on characteristics that they are too afraid to enact outside of the game and permits them to question their own personalities and the places in which they find themselves. They find a free space to take on other roles and repeat them to perfection. As Forster points out, in the couple's conversation, there are constant allusions to a script in these games that suggests a certain theatricality to their role-playing.[58] Here, the emphasis on games allows us to examine more closely the use of circularity in *Dos viejos pánicos* and its purpose in the larger context. In the following scene, Tabo and Tota discuss their game and Tabo suggests a change:

> *Tabo*: Estaba pensando que en la próxima sesión de juego podemos hacer que Tota mate a Tabo y lo entierre en el hoyo.
> *Tota*: Por poder hacerlo no quedará... Ya sabes que cuando uno está muerto puede hacer lo que quiera. Pero no sé, así de pronto la idea no la veo bien. ¿Qué te propones?
> *Tabo*: Bueno, como proponerme, nada. Es tan sólo una variación dentro del juego. En la primera, Tota mata a Tabo; en la segunda, Tabo mata a Tota.
> *Tota*: No me parece correcto disponer de Tota y de Tabo para que Tota mate a Tabo o Tabo a Tota. Ellos están muertos de verdad y ya no podrían matar. Otra cosa sería si estando vivos aceptaran jugar al juego.
> *Tabo*: Pero, a lo mejor, Tabo y Tota... (*Se calla*.)
> *Tota*: (*Se para, va a la cama de Tabo.*) ¿Quieres decir que Tabo y Tota se quedaron con las ganas de matar?
> *Tabo*: Con las ganas de matar, exactamente, Tabo a Tota o Tota a Tabo.
> *Tota*: ¿Tú quieres decir que también ellos tuvieron miedo?
> *Tabo*: A lo mejor... Porque, oye, uno ve a la gente y piensa: ese es un tipo de pelo en pecho o esa es una tipa que se lleva al más pintado por delante, y un buen día te enteras de que son unos cobardones. O no te enteras, pero en el fondo lo son.
> [*Tabo*: I was thinking that in the next session of the game we could make Tota kill Tabo and bury him in a hole.
> *Tota*: Of course we could do it... You know that when one is dead one can do what one wants. But I don't know, I'm not sure if I like the idea. What are you thinking?
> *Tabo*: Well, thinking, nothing. It's just a variation on the game. In the first, Tota kills Tabo; in the second, Tabo kills Tota.
> *Tota*: I don't think it's right to have Tota and Tabo at our disposal so that Tota kills Tabo or Tabo Tota. They're really dead and they couldn't kill anymore. It would be another thing if they were alive and they accepted playing the game.
> *Tabo*: But, maybe, Tabo and Tota... (*He goes quiet*.)
> *Tota*: (*She stops and goes to Tabo's bed*.) Are you saying that Tabo and Tota still want to kill?

Tabo: Want to kill, exactly, Tabo Tota or Tota Tabo.
Tota: Are you saying that they are also afraid?
Tabo: Maybe...Because, hey, you see people and you think: that guy is a real macho or that's a girl who could take on anyone, and then one day you find out they're cowards. Or you don't find out, but in the end they are] (Act 2: 492–3).

This scene offers us the opportunity to examine the mechanics of the couple's game. Here, Tabo and Tota discuss some possible alterations to their game of "killing" one another—each of their self-created performances, though using the same script, allows for improvisation and innovation. It is interesting to note that both Tabo and Tota refer to themselves in the third person, suggesting a disconnection at this moment between their two selves—the "real" Tabo and Tota and those of the game. By discussing themselves as entities outside of the conversation, they detach these selves from the emotions that are being discussed and they emancipate the Tota and Tabo that are discussing the game from those who are playing it. In this way, they gain a distance that allows them freedom and courage to act upon their own thoughts and wishes. The couple sees themselves as playing a role in a game that they have created and in which they repeatedly participate.

The idea of repetition is particularly interesting in relation to Arrabal's Panic Theater, where it has been explored in depth.[59] Circularity and repetition in Cuban theater (including Piñera's *Una caja de zapatos vacía* (1968) and Triana's *La noche de los asesinos*) create a spectacle that allows both the playwright and the spectator to consider the further-reaching goals of the play and the social context in which it is written.[60] This theater is focusing on the lack of a break with the past and the circularity found in the political and social arena and, in this way, invites the spectator to question to what extent the society has advanced.[61] Piñera's *Una caja de zapatos vacía* allows the reader to examine in even more detail the degenerative effects of repetition through the re-birth of one of its characters from the body of another. Here, rebirth becomes the sign of a persistent repetition that never allows for change or renovation.[62] These observations are important when considering Piñera's *Dos viejos pánicos*, since inherent in the idea of revolution, there is a double violence: that of the break with the past and that of returning incessantly over the same events.

Virgilio Piñera uses the tactic of a game being repeated over and over in *Dos viejos pánicos* in order to focus on the social stagnation he is encountering. As seen in their conversations, Tota and Tabo are engaging in an activity that they repeat and fine-tune daily, but that shows no benefit to them personally or socially. Nevertheless, their role-playing affords them a brief respite from the fears of everyday existence, though, as George Woodyard points out, this is not a way out from circularity: "They stay alive in order to play dead, and playing dead enables them to stay alive. [...] In 'death' they experience the same emotions and the same contradictory relationships which exist in life."[63] For both the Absurdists and the Panics, circularity is just an aspect of life that shows its irrationality and lack of meaning. For Piñera, in turn, it seems to suggest a need to break from fear and liberate ourselves from its oppression.

In a turn away from their earlier fighting of one another, Tabo and Tota begin to collaborate when their attention turns to an outside authority. This collaboration begins in earnest when they move to the task of filling out an official *planilla* that they have been asked to do. With the introduction of the *planilla* in *Dos viejos pánicos*, an external power enters into the play and we can observe what may possibly happen in an imaginary confrontation between the couple and the authorities they seem to fear so intensely. The confrontation begins when Tabo remarks that Tota's indigestion comes from the fact that a man will come to pick up the *planilla* the following day. She reacts with fear and states that she becomes more and more afraid every time she reads the questions. Within this fear, one can perceive common anxieties expressed when faced with any official paperwork and the spectator can sympathize with the difficulties of bureaucratic red tape. Nevertheless, when we hear the questions, this *planilla* seems to hide a more sinister motive. While Tabo reads aloud Tota's questions, they come to the decision that the *planilla* reveals the immense knowledge that the questioner has about their entire lives:

> *Tabo*: ¿Te das cuenta? Saben tanto de ti y de tu vida que son como si fueras tú misma.
> *Tota*: Son preguntas que meten miedo. ¿Cómo contesto ésta?
> *Tabo*: Lo que contestes no va a tener mayor importancia. Fíjate, en el fondo ellos no preguntan.
> *Tota*: (*Gritando*.) Y si no preguntan, ¿qué carajo es lo que hacen?
> *Tabo*: (*Con mucha calma*.) Contestan, contestan por ti. (*Pausa. Pasa la vista por la planilla*.) Ya te dije que la respuesta de cada pregunta está en la pregunta siguiente. Así, la pregunta número uno es contestada por la número dos, ésta por la tres, la tres por la cuatro y la cuatro por la cinco.
> [*Tabo*: Do you realize? They know so much about you and your life that it's as if they were you.
> *Tota*: The questions are scary. How do I answer this one?
> *Tabo*: Whatever you answer doesn't matter. Look, in the end, they're not asking.
> *Tota*: (*Yelling*.) If they're not asking, what the hell are they doing?

Tabo: (*Very calmly.*) They're answering, they're answering for you. (*Pause.* He looks at the planilla.) I already told you the answer for each question is in the following question. So, question number one is answered by number two, that one by three, three by four, and four by five] (Act 2: 495).

Tabo and Tota come to the conclusion that the *planilla*'s purpose is to reveal to the couple how much the *planilleros* know and, thus, to show their incessant power over Tabo and Tota. In this way, the title's *pánicos* are exposed even further. The intimate details of Tabo's and Tota's lives—their previous loves and their reasons for marrying one another—are uncovered in a simple, official *planilla* that they are to fill out. The realization that the outside other knows more about them provokes fear in the couple. The fact that the *planilla* is not questioning but unveiling the depths of official knowledge confirms the very fears, or panic-stricken thoughts, that they have been professing throughout the play. Matías Montes Huidobro in *Persona, vida y mascara en el teatro cubano* sees the *planilla* as the introduction of "un elemento totalmente absurdo [a totally absurd element]" in that it becomes an all-encompassing cycle that is impossible to escape.[64] They become stuck in a circular trap and the fact that the *planilla* answers its own questions shows its lack of meaning and, I would argue, gives it a meaning that goes beyond the questions and answers. It is this meaning that frightens Tota. Hidden in this meaning, we can see the more sinister implications of an official body that knows everything about its subjects, an allusion that can be seen to point to the Revolutionary government and the increased controls that became a part of everyday life in the 1960s.

When considering the importance of the *planilla* within the play, we should turn once again to Arrabal and his Panic Theatre in order to understand its role in *Dos viejos pánicos*. For Arrabal, objects within the context of the theater were regarded as vital to the play's direction. In fact, Arrabal imagined a theater where the objects would become autonomous and replace the characters.[65] As Francisco Torres Monreal points out in reference to Arrabal's use of theatrical objects: "Dentro del teatro moderno, Arrabal es posiblemente el dramaturgo que mejor intuye las enormes posibilidades poéticas del objeto teatro cuyas funciones primordiales seguían siendo, en la trayectoria realista, las de figurar su papel utilitario o descriptivo de un ambiente o personaje [In modern theater, Arrabal is possibly the playwright who best senses the enormous poetic possibilities of the object theater whose primordial functions continue to be, in the realist trajectory, those of representing the utilitarian or descriptive role of an atmosphere or character]."[66] For Arrabal, then, the objects that are used within the context of a play are not chosen arbitrarily, but are deliberately placed within the play's argument in order to signify something more. Thus, if we are going to recognize a connection between Arrabal's Panic Theatre and Piñera's text, we should examine closely the role that the *planilla* occupies in this scene and in the play more generally.

It can be easily understood that, for Tabo and Tota, the *planilla* is more than a simple form that they need to fill out. In *Dos viejos pánicos*, it takes on a symbolic role that is similar in some ways to that of Fear. With the *planilla*, the outside creeps into and contaminates the inside world that Tabo and Tota have painstakingly created for themselves. Everything that they fear outside becomes personified in this questionnaire. By inspiring fear in the characters, the *planilla* becomes the materialization of that fear within the scene. For Arrabal, theatrical objects acquire a symbolic and multivalent interpretation. As Torres Monreal explains, "la mayor parte de los objetos arrabalianos se integran en un mundo de relaciones dentro de la ceremonia que justifican su interpretación simbólica (y por ello mismo plurivalente), por más que el dramaturgo insista en su carácter realista puramente denotativo [the majority of Arrabalian objects are integrated into a world of relations in a ceremony that justifies their symbolic interpretation (and for that reason polyvalent), no matter how much the playwright insists on the purely denotative realist character]."[67] What is important is to understand the role that the *planilla* is meant to occupy within the play—its symbolic meaning. Primarily, the *planilla* refers to official authority—that which Tota and Tabo fear the most. In the couple's conversations, this "authority" takes on the role of "la oficina de preguntas [the office of questions]" and the most important of all, "el jefe de los planilleros [the boss of the *planilleros*]," although I believe that it reaches beyond this single being to encompass all authority figures associated with what can be thought of as "official" business (Act 2, 494, 495). That is, the *planilla* represents all of the offices before which Tota and Tabo, and all of those who can identify with their fear of this official bureaucracy, must yield. In this way, the *planilla* becomes a representation of any figure that can exercise power and control over Tota and Tabo. Thus, when they deduce that the *planilla* is not seeking answers from them but responding to its own questions through its vast knowledge of their life, it becomes another thing to fear—another *pánico* that occupies their lives. The *planilla* symbolizes their fear.

The culmination of Tabo and Tota's struggle with Fear is located towards the end of the play, when the couple attempts to kill their eternal enemy, the abstract Fear. The fact that the *planilla* answers its own questions naturally sparks even more fear and violence among the two, given that there seems to be no way out. As a result, they turn violent against one another. In the physical fight that follows, once again they "kill" one another and pass into a realm where fear does not exist, and they are again capable of actions unimaginable under normal circumstances:

> *Tota*: [...] Ya estoy muerta. Y también tú. ¿No te das cuenta de que estamos muertos?
> *Tabo*: ¿Estamos muertos, Tota? ¿De verdad que lo estamos?
> *Tota*: Te lo juro. A ver, ¿sigues teniendo miedo?
> *Tabo*: (*Sacando las piernas de encima de Tota, se tiende al lado de ella.*) Ni pizca de miedo. (*Pausa.*) ¡Qué bien se está así, a tu lado! Ahora podemos hacer todo lo que nos salga de adentro. Por ejemplo, podemos romper las planillas. (*Las saca del bolsillo.*) Toma la tuya. (*Se la pone en las manos.*) Rómpela. (*Rompe la suya.*)
> *Tota*: (*Rompe la planilla.*) ¡Qué felicidad! Es como nacer de nuevo, sin culpa ni pena. Siempre estamos discutiendo y tirándonos los platos a la cabeza, pero al final nos ponemos de acuerdo.
> *Tabo*: De acuerdo, eso es, siempre de acuerdo. (*La acaricia.*) Mi Tota, mi Totica, qué hubiera sido mi vida sin ti.
> *Tota*: ¡Y la mía! (*Lo acaricia.*) Mi Tabo, mi Tabito. (*Pausa.*) Me siento tan feliz que pienso seguir muerta hasta que me muera.
> *Tabo*: ¡Qué magnifica idea, Tota, qué magnifica idea! Seguir muertos hasta que nos toque morirnos. Mi vieja, eso es un descubrimiento sensacional. (*Pausa.*) ¡En esta casa se acabó el miedo! (*Coge a Tota por un brazo.*) ¡Ven, vamos a matarlo!
>
> [*Tota*: [...] I'm already dead. And you are too. Don't you realize we're dead?
> *Tabo*: We're dead, Tota? We really are?
> *Tota*: I swear. Let's see, are you still afraid?
> *Tabo*: (*Taking his legs off Tota, he lays down next to her.*) Not afraid at all. (*Pause.*) It's so nice here, by your side! Now we can do everything that we want. For example, we can rip up the *planillas*. (*He takes them out of his pocket.*) Take yours. (*He puts it in her hands.*) Rip it. (*He rips his.*)
> *Tota*: (*She rips the* planilla.) I'm so happy! It's like being born again, without guilt or pain. We're always arguing or throwing plates at one another's heads, but in the end we agree.
> *Tabo*: Agree, that's it, always agree. (*He caresses her.*) My Tota, my Totica, what would my life be without you.
> *Tota*: And mine! (*She caresses him.*) My Tabo, my Tabito. (*Pause.*) I feel so happy that I think I'm going to stay dead until I die.
> *Tabo*: What a magnificent idea, Tota, a magnificent idea! Stay dead until we die. Old girl, that's a wonderful discovery. (*Pause.*) In this house fear is dead! (*He grabs Tota by the arm.*) Come on, let's kill it!] (Act 2: 496–497).

In this scene, we can observe the change in the couple's relationship from antagonists in a physical struggle that brings about their own "end" to a couple working together against a common enemy. Upon observing that they are no longer afraid, now that they are dead, Tabo and Tota rip up the *planilla* and, thus, liberate themselves from their past.[68] The omniscient knowledge that the *planilla* exhibits at this point does not provoke fear, but defiance; they can now destroy not just the *planilla* but all that it represents. We return to Tota's earlier observations that death brings about a life free of consequences, which, in turn, signals an existence free of fear. Curiously, this freedom from fear provokes in the two a desire to kill once again, though their new victim is fear itself—a character that would seem to have disappeared after they died. Thus, the couple's "death" at the hands of the other—a death provoked by their own fear and feelings of helplessness, but that also liberates them from that fear and helplessness—provides them with the opportunity to come together again as a couple and with the ability to transcend their fears and to do away with that which tyrannizes the two. Furthermore, this desire to kill Fear is a radical move that indicates their longing to regain control over their own lives. Implicit in this is a criticism against the official *planilleros* who have driven them to such extremes through mechanisms of control and repression.

Tabo and Tota track down what they believe to be Fear to one of the beds that is on the stage. After a struggle, in which Tabo represents the dialogue of the imagined Fear dying at Tota's hands, the two briefly celebrate before they take away the pillow to discover that once again Fear has eluded them and escaped. Meanwhile, during the previous struggle on the bed, according to the play's stage directions, the stage light had been shrinking until it was nothing more than a small cone of white light that appears in the center of the scene, the size of a basketball. While the couple attempts to find Fear underneath the bed and the mattress, the small cone of light, that now symbolizes Fear, begins to move across the stage until it is only a meter away from the couple:

> *Tabo y Tota buscan alrededor de las camas haciendo gestos muy estereotipados de consternación mientras repiten "<<se fantasmó>>," en un tal crescendo que, finalmente agotados, caen de rodillas con sus cabezas descansando sobre el piso. En dicha postura se mantendrán un segundo. Entretanto el cono de luz se habrá movido para colocarse a solo un metro de ellos*

[Tabo and Tota look around the beds making stereotypical gestures of dismay while they repeat "it disappeared," in such a crescendo that, finally worn out, they fall to their knees with their heads resting on the floor. They stay in this position for a second. Meanwhile the cone of light has moved to within a meter of them] (Act 2: 502).

Despite all of their desires and effort, Fear proves to be slippery and eludes them again. The movement of the light across the stage subtlely, but firmly, reminds them of this. Nevertheless, Tota discovers the cone of light and the two put on their bravado and try once more to capture it, projecting their own fear onto the cone of light:

> Tota: (*Se agacha y cuando va a poner sus manos sobre el cono de luz, éste salta y se fija en la pared lateral izquierda. Lanza una carcajada, se dobla de la risa.*) Míralo, Tabo; está que se caga del susto. Vivir para ver. Así que el machazo de la película...Ja, ja, ja...Anda, baja y métenos miedo. Vuelve a hacer todo lo que has hecho en tu puñetera vida. (*Pausa, hace pabellón con la oreja.*) ¿Qué...? ¿Qué...? Eso quisieras. No, el que está cagado del susto eres tú. (*Pausa, se vuelve hacia Tabo.*) Óyelo, dice que estamos temblando. Dile algo, Tabo, dile algo con tu voz de capitán intrépido.
>
> [*Tota*: (*She kneels down and when she's about to put her hands on the cone of light, it jumps away and moves to the left hand wall. She doubles over from laughter.*) Look at it, Tabo; it's scared shitless. Live to see the day. So, the big man of the movie...Ha, ha, ha...Come on down and scare us. Repeat everything you've done in your damn life. (*Pause, she cups her ear.*) What...? What...? You'd like that. No, the one who's scared shitless is you. (*Pause, she turns toward Tabo.*) Listen to it, it says we're shaking. Tell it something, Tabo, tell it something with your fearless captain voice] (Act 2: 502).

In this scene, the couple, who violently attacked one another in the previous act, now work for a common goal. Tota's words intend to taunt Fear and show her own domination of the new circumstances that she and Tabo have brought about. Instead of inspiring fear, the elusiveness of the light only provokes contempt from the couple. The earlier act of ripping up the *planilla* has emboldened the two to take on the very monster that haunts their existence. Yet what is even more amazing is the cooperation that they are initiating with this new undertaking. It indicates a collaboration that negates the division between them that we saw before and suggests a possible return to humanity and the cooperation that begins any union of love.

The optimism and bravado that can be seen here continues when the cone of light "se posa en el pecho de Tabo [rests on Tabo's chest]" which is followed rapidly by another on Tota's chest (Act 2: 503). In order to rid themselves of this incarnation of fear (embodied here by light[69]), the couple tries to asphyxiate the two cones of light between their bodies by hugging one another fiercely, contrasting strongly with their earlier abuse of one another:

> Tabo: (*Se abraza a Tota.*) Por ti.
> Tota: (*Se aprieta aún más a Tabo.*) Por mí.
> Tabo: (*Se aprieta aún más a Tota.*) Por ti.

Tota: (*Se aprieta aún más.*) Por mí.
 De tanto apretarse uno contra el otro en medio de un atroz jadear, caen al suelo lentamente. Ahora el cono de luz ha aumentado su tamaño y les cubre enteramente el cuerpo.
[*Tabo*: (*He hugs Tota.*) For you.
Tota: (*She hugs Tabo even more.*) For me.
Tabo: (*He hugs Tota even more.*) For you.
Tota: (*She hugs Tabo even more.*) For me.
 From so much hugging one another, while panting violently, they slowly fall to the floor. Now the cone of light has grown in size and entirely covers their bodies] (Act 2: 503).

This scene represents the most human moment of the play, where the two characters collaborate in order to look for a possible exit from the circularity of fear. It promises that, through their physical caresses and the manifestations of their love, they will find an eternal hope that will erase their daily fights and insults and, in the process, kill that which haunts their existence. In this way, the spectator can see the possibility of a way out of fear through collaboration and love, two traits that have been absent from the play. Fear, represented with a single cone of light that ends up eclipsing the two, overcomes them and they collapse to the floor despite the ever-increasing strength of their amorous demonstration. With this action, they see the impossibility of killing Fear, which only grows the harder they try.

Following this final defeat, Tabo and Tota briefly and horrifically return to their daily routine, as is illustrated in the scene quoted below. However, I believe that this battle against the elusive Fear that exists in their lives is symbolic of the small victories and large defeats that characterize Tabo and Tota's existence. The fact that they came together to collaborate against Fear and the repetition that is inherent in their "killing" game hints at a future, although distant, success in their daily battle.

Virgilio Piñera's *Dos viejos pánicos* closes with the end of Tabo and Tota's day and this underscores their future return to the same daily routine. The spectator can, thus, see both the disappointment of defeat and the security of routine. The repetition of their daily schedule and, also of words in the following quote, underlines their inability to break free from their fear and captivity. At the same time, the memory of their recent collaboration against Fear and the renewal of their relationship born from their shared struggle suggest that tomorrow may bring about their long-awaited success:

Tabo: Tota, ¿qué vamos a comer mañana?
Tota: Carne con miedo, mi amor, carne con miedo.
Tabo: ¿Otra vez? Ya no la resisto.

Tota: ¿No la resistes, de verdad que no? Pues entonces comeremos miedo con carne. (*Pausa*.) Y ahora, duerme, mi amor. Hasta mañana.
Tabo: Hasta mañana. (*Pausa*.) Tota...
Tota: ¿Qué?
Tabo: ¿Mañana será otro día?
Tota: Sí, Tabo, otro día, otro día más...
Tabo: (*Suspira*.) Otro día más...
Tota: Y otra noche más...
Tabo: Y otro día más...
Tota: Y otra noche más...
Tabo: Y otra noche más y otro día más...
Tota: Y otro día más y otra noche más...
    Cuando Tabo dice "Otro día más," el telón empezará a cerrarse muy lentamente.

[*Tabo*: Tota, what are we going to eat tomorrow?
Tota: Meat with fear, my love, meat with fear.
Tabo: Again? I can't do it.
Tota: You can't, really you can't? Well, then, we'll eat fear with meat. (*Pause*.) And now, sleep, my love. See you tomorrow.
Tabo: See you tomorrow. (*Pause*.) Tota...
Tota: What?
Tabo: Tomorrow will be another day?
Tota: Yes, Tabo, another day, another day more...
Tabo: (*He breathes*.) Another more day...
Tota: And another night...
Tabo: And another day...
Tota: And another night...
Tabo: And another night and another day...
Tota: And another day and another night...
    When Tabo says "Another day," the curtain begins to slowly close] (Act 2: 509).

While this final moment in the play begins with the reminder of their fear and its certain return tomorrow, Tota and Tabo recognize that tomorrow is also another day where the possibility to break free from fear exists. The quotidian nature of Fear—that it is a daily companion along with their meals—offers the possibility that they may break free from it and the inevitability of a repetition of their murderous encounter with it precisely because of their daily practice routine. Repeating every day their own existence free of fear and the murder of fear implies that someday the dress rehearsals will end and they will be ready to carry through with their actions. This repetitive routine that they reveal in their final exchange uncovers both a pessimism that nothing ever changes and an optimism that some day it may.

However, this optimism is quickly contradicted by the fate of Piñera's play, given that rather than optimism and change, it faced censorship. It is important at the end of this chapter to return to the earlier discussion of censorship, specifically that of the play's representation in Cuba at the moment of its writing. It has been established that the objective of much of the theater written and produced in the 1960s and 1970s, and in general, is to engage with its audience and, in that way, focus attention on important issues and problems of the moment. *Dos viejos pánicos*, like some of the other plays examined here, was not represented in the context in which it was written until much later. Thus, the relationship that the playwright imagined when formulating his play is never established and the conversation that is initiated is suspended. This was surely due to the controversial material, the absurd nature of the play, or the fact that its author was a gay man in a homophobic time. This fate reminds us of the writings of Rene Girard discussed in the Introduction on the indispensable role of violence to a community. Girard believed that violence was inevitable and a ritual violence would allow a society to control these acts.[70] In this way, we can consider the awarding and subsequent silence of Piñera's *Dos viejos pánicos* to be a ritual bloodletting that would stem the tide of an all-encompassing violence. Similarly, it served as a negative example of what would not be tolerated.

However, in a twist that reveals a certain optimism, despite the lack of a Cuban premiere, it appears that *Dos viejos pánicos* was read publicly in Havana. This does suggest the beginnings of a conversation on fear and the issues explored in *Dos viejos pánicos* in the 1960s in Havana but similarly offers a truncated view of Piñera's initial desire when writing his play. While a reading of a play does begin to introduce it and its topics to an audience, it lacks the depth and exploration that a true dramatic production has. In this way, Piñera's *Dos viejos pánicos* was never able to culminate in a production that would couple the playwright's words with the actors' gestures and with the audience's impressions. The play, then, becomes a different entity that seems to be suspended between the vision of a single person and the collective production that will continue to question and provoke as long as it is represented and seen.

The topics of *Dos viejos pánicos* go much further than Virgilio Piñera's possible political affiliations, even when many of his critics have been tempted to read coded political commentaries of the particular Cuban situation in the 1960s. While we can see allusions to the increased control that marked the 1960s under Castro in Cuba, there are also more general references that make Piñera's work go beyond the geographical and temporal contexts in which he is writing. Beyond any of these specific readings, it is clear that Piñera wants to provoke a discussion on the universality of fear—that we all have profound fears—and the violence that results from our desire to hide our fear. Concretely, however, Piñera draws from his own national context and the spectacle that was unfolding before him to illustrate the inescapability of fear. The violent spectacle resulting from the fear that Tota and Tabo create alludes to the political and social context of Havana and Cuba in the 1960s and shows the circularity of fear and violence. Their violence produces not just violent acts, but at the same time a complete circuit of violence that, in turn, includes a violent repression of the self. Thus, there does not seem to be any exit to violence but through its violent repression or violent expression itself, which can be observed in the relationship between Tabo and Tota.

Facing violence, there is a certain circularity that can culminate in a dead-end road. It is in this way (in the exit from violence) that Piñera, nevertheless, shows his particular humanism. The characters are not trapped beings in an existentialist labyrinth, but in the end, beings that find in love, in collaboration, a possible redemption—albeit ironic—from the violent cycle. In this way, the play discusses the possibilities of human beings and facilitates a self-criticism of the potential that we all hide. Remembering Artaud's assertions and his theater of cruelty along with those of Albuquerque in reference to the Latin American context, *Dos viejos pánicos* tries to force the spectator to reflect on the presented topics and the context that are manifested. Nevertheless, Virgilio Piñera turns the gaze of his spectators to the internal situation and the eternal struggles that we all daily fight, causing a self-examination that will promote a breaking free from this circularity. As one of Piñera's most important theatrical works and given where it falls chronologically, *Dos viejos pánicos* locates violence first as domestic and quotidian and as a result of an external source that threatens the characters. Over the course of the next few years, other forms of violence would be represented dramatically, such as those related to historical, identity, and spectatorship. In the next chapter, the topics of revolution, censorship and the past will continue to be important in the analysis of *La dolorosa historia del amor secreto de don José Jacinto Milanés* by Abelardo Estorino. Here we move into the early 1970s in Havana, a time that became more and more restricted in artistic exploration as the official government tried to narrow definitions of what it meant to be revolutionary. The way these definitions clash with the goals of the theater can already be seen in *Dos viejos pánicos* but will continue in Estorino's *La dolorosa historia* through the exploration of the writing of history. Nevertheless, as the first play studied here, *Dos viejos pánicos* reminds us that the beginning of violence is that of the self and the other.

Chapter 2

Cobwebs of Memory: History Made with Violence in Abelardo Estorino's *La dolorosa historia del amor secreto de don José Jacinto Milanés* (1974)

> ¿En qué otro país del mundo hay una provincia que se llama Matanzas?
> Guillermo Cabrera Infante, Vista del amanecer en el trópico

In *La dolorosa historia del amor secreto de don José Jacinto Milanés* (1974), Abelardo Estorino (b. 1925) draws on the life of a nineteenth century Cuban poet and playwright and explores the correlations between that time period and Estorino's own in a way that reveals the betrayal, violence and pain that typify Cuba's literary community in the 1970s. *La dolorosa historia*, like *Dos viejos pánicos*, has an uncertain relationship with stage production. It was written in 1974 though it did not premiere until 1985, staged then by the theater group *Teatro Irrumpe* directed by Roberto Blanco. It is a play that examines the role of violence in the making of history and memory through the life of Milanés, a figure that allows Estorino to enter into the past in order to consider history, memory, and betrayal onstage and simultaneously to allude to the present historical and social moment. The spectacality of the violence that surrounds Milanés' life within *La dolorosa historia* refers to the (literally) spectacular violence that was being mounted in the cultural and political context of Cuba in the early 1970s, as was shown in the introduction with the *caso Padilla*. Just as Piñera explored the prevalence of fear, Estorino's play affords the playwright a space in which his authorial hand guides the characters and points to the political events that shaped the 1970s through the past. This historical context is one that includes episodes discussed in reference to *Dos viejos pánicos*, such as the definition of the Revolution, the infamous *caso Padilla* and the censorship that characterized Cuba in the late 1960s and the 1970s and, thus, highlights the role that spectacle played in the contemporary context through the exploration of Milanés' life and circumstances.

La dolorosa historia is divided into seven scenes that revisit an important moment of the poet and playwright José Jacinto Milanés' life. The final scene, "Delirio [Delirium]," which retells his loss of reason, serves as the center of the play's objectives, by mixing Milanés' interior life with the uprising known as the *Conspiración de la Escalera* [Conspiracy of the Ladder] in Matanzas in 1844 to question race relations in nineteenth-century Cuba and identify a parallel between the private and the public. The violence depicted on stage in *La dolorosa historia* is not limited to that of the historical rebellion but alludes to the social violence inherent in slavery and racism. Simultaneously, Milanés' questioning of the belief that violence is needed to purge the past sins recalls the context of the Cuban Revolution where a new era violently ended an earlier one. The focus in Estorino's play breaks with a historical circularity that he identifies in the two moments and reminds the spectator of the stronger regulations around literature in Cuba that came into effect in 1971 with the Congress on Education and Culture. By focusing on the use of violence in the portrayal of an uprising for social equity in Cuba in the final scene, Estorino's play promotes a renewed discussion of the Cuban Revolution through nineteenth-century historical events. Inherent in this temporal parallel, there is a decided effort to change the canonical viewpoint through a questioning of the interpretation of past events.

History plays a vital role in *La dolorosa historia* in reference to the life of José Jacinto Milanés and the nineteenth-century context in which he lived, as well as that of the immediate past of the 1960s and 1970s in which Abelardo Estorino writes the play. For this reason, it is necessary to explore these two distinct historical moments to understand their larger meaning within the play. Cuba's nineteenth century is full of war and injustice. While many of Spain's colonies in the Americas had gained their independence, Cuba remained in the hands of the colonial government in Madrid throughout the nineteenth century. The initiative of the Haitian slave rebellion at the end of the eighteenth century as well as Spain's renewed commitment to maintaining control of Cuba contributed to its continuation as a colony.[1] In the Atlantic world of the nineteenth-century, abolition of slavery was a principal topic of discussion, and the Cuban political and social elite was no exception. In 1808, the slave trade with British ports was outlawed and the British government began to pressure Spain to quickly follow its example. In 1820, in exchange for a payment of 400,000 pounds to compensate the economic loss, the slave trade in Spain was abolished. However, it continued illegally for another fifty years, given Spain's reluctance to take a stand against the Cuban plantation owners and the latter's fervent desire to continue farming their tobacco and sugar plantations with slaves.[2] With this, the scene was set for constant friction between Cuba and Spain around both the topic of slavery and of independence, and these debates persisted on the island, as can be seen in the figure of José Jacinto Milanés.

On the local level, the Conspiracy of the Ladder provided the immediate backdrop for Estorino's play. Indeed, the final scene of *La dolorosa historia* enacts the execution of the poet Plácido, one of the defining moments of the Conspiracy. The Conspiracy, taking place in Matanzas in 1844, was an uprising of slaves, free blacks and white abolitionists in order to end slavery and obtain independence. It was discovered and the supposed planners were arrested. The free blacks and slaves were accused of conspiring against the government and tied to ladders where they were tortured until they confessed or died. The whites were arrested or fled into exile. This period of repression lasted for about six months and almost two hundred slaves and free blacks were shot or tortured to death on the ladder. As a result, life became even more difficult for slaves and free blacks in Matanzas and Cuba, more generally. Many plantation owners or prominent whites that were seen to be sympathetic to or involved in the Conspiracy were exiled. However, since the very days of the Conspiracy, there have been questions over the authenticity of the events: whether there was an actual uprising being planned or whether it was a plot to vilify the free blacks and the abolitionist and independence movements.[3]

Estorino wrote *La dolorosa historia* in 1974, an important moment that positions the play to evaluate the legacy of the Cuban revolution and assess the contemporary moment. The beginning of the 1970s was a pivotal time in the Revolution. In 1970, sugar, always a central issue in Cuban social and political circles, given its centrality to economic success, came to take center stage. The harvest of this year, known in Spanish as the *Zafra de los diez millones* [Ten Million Ton Harvest], became a very public campaign to harvest ten million tons of sugar. This was to be by far the biggest harvest ever, thus providing a moral victory for the Revolution at a time when it was being questioned at home. However, it fell short of its goal, and the government was forced to shift its economic policies to ones that had more realistic objectives, but that never garnered the same nationalistic fervor of the earlier effort. This public failure helped to set the tone for the change in political and social context that would characterize the early 1970s in Cuba.

On the political front, the year 1971, witnessed the culmination of the *caso Padilla* that exploded both nationally and internationally. As discussed in the Introduction, this episode dated from the 1968 awarding of a prestigious literary prize for Antón Arrufat's play *Los siete contra Tebas* and Heberto Padilla's collection of poems *Fuera del juego*, despite their supposed "counterrevolutionary" ideology. Both authors suffered censorship and alienation from Havana's intellectual circles: Arrufat was relegated to working in a municipal library, exiled from the theater world, while Padilla was incarcerated in 1971 and forced to read a mea culpa. In a public spectacle that paralleled much of what was happening onstage, this imprisonment sparked an international condemnation of the Cuban Revolution by such staunch supporters as Jean Paul Sartre, Octavio Paz and other international writers and artists.

Many of the social and political pressures from these events came to a head in the Congress on Education and Culture in 1971, where new, harsh regulations were enacted in order to control and monitor the university and artistic communities. Publication standards were created to dictate the essential revolutionary quality of all works published or prized in Cuba.[4] As a result, the Casa de las Américas began to award works that were clearly political and revolutionary in nature, breaking from the prizes awarded in the 1960s when many of the winners were not openly political in the same way.[5] This event is just one of many that took place during what has come to be known as the *quinquenio gris*, a period of five years in the early 1970s (1971–1976) known for repression and persecution. Although *La dolorosa historia* was not awarded a Casa de las Américas prize for the year it was published, it is important to consider this policy change when analyzing the play. While Estorino had generally written plays that used a realist portrayal of its subject matter and had thus not seemed to be directly affected by the new regulations, he chose to indirectly discuss them in *La dolorosa historia* through various strategies that will be explored later.

The term *quinquenio gris*, translated as Gray Years or Gray Half-Decade, is particularly interesting, not just for its coining, but also for the fact that there has been a recent return to it in the last few years. *Quinquenio gris* was first used by Ambrosio Fornet, a coining that he discusses in his recent "El Quinquenio Gris: Revisitando el término [The Gray Years: Revisiting the Term]." This was a paper that formed a part of the January 2007 conference titled "La política cultural del período revolucionario: Memoria y reflexión [The Cultural Politics of the Revolutionary Period: Memory and Reflection]," organized by the Casa de las Américas. This conference is part of the recent reflections on the past that have set Raúl Castro's tenure apart from his brother's. In this essay, Fornet offers a detailed discussion of the atmosphere in the 1960s, leading up to the first half of the 1970s. In the final sections, he turns his attention to the years between 1971 and 1976, the time now known as the *quinqenio gris*. This, however, is only one of the terms used to refer to these years, another being the *Pavonato*, in reference to Luis Pavón Tamayo. Pavón was the head of the National Council of Culture from 1971 to 1976 and was responsible for implementing many of the policies that would define these years as repressive. Pavón was, also, reportedly, the real identity of Leopoldo Ávila, the pseudonym for a series of articles published in *Verde Olivo* beginning in November of 1968.[6] *Verde Olivo* was the magazine of the *Fuerzas Armadas Revolucionarias* in Cuba. These articles were written against the actions and writings of Guillermo Cabrera Infante, Heberto Padilla, and Antón Arrufat.

It is important to consider the recent reflections on this time period and terminology that have characterized the twenty-first century and particularly Raúl Castro's government, given that it reveals a desire to examine the past to make the present and the future more open. To this end, Casa de las Américas held the conference in January of 2007 of which Ambrosio Fornet formed a part. Attending and participating were important intellectuals and members of the artistic community from both the early years of the Revolution and those currently active in these communities. This mix of first-hand participants and the important current voices on Cuban culture on the island shows a desire to revisit and explore the past. Nevertheless, this action may also remind the cynical spectator of the Revolution's repeated "rehabilitation" of certain figures from the artistic communities after their deaths, Virgilio Piñera being perhaps the best example.

Abelardo Estorino, who was a witness to these events, was born in 1925 in the province of Matanzas. He studied to become and, then practiced as, a dentist before joining the Teatro Estudio group in 1960. He is most famous for his play *El robo del cochino* [*The Theft of the Pig*] written in 1961, which is often considered one of the masterpieces of the Revolution, given that it portrays how the theft of a pig by a poor youth breaks apart a father and son, revealing the father's bourgeois ideals and the son's nascent commitment to social justice. Since this play, Estorino's theater has held a central place on Cuban stages. His plays have typically been identified as examples of realism, given that they tend to explore topics much more associated with the quotidian. *Los mangos de Caín* (1967) is Estorino's one exception to realistic theater. This is a play clearly allegorical in meaning and often rejected by critics.[7] Estorino often focuses on familial issues in the Cuban context, as is the case in *El robo del cochino*—as well as questions about the revolution and its ongoing commitment. His realist view of Cuban society permits his works to examine central issues that are often ignored, as we see with his dramatic biography of the nineteenth-century poet and playwright José Jacinto Milanés.

José Jacinto Milanés, a fascinating figure for Estorino and for other writers in the 1960s, was born in Matanzas (Estorino was also born in the province of Matanzas) in 1814, the first son of a well-respected family who had more children than money to support them.[8] Estorino's play follows Milanés' life closely, focusing on particular moments that concern the play's overarching themes. Milanés' family's lack of means impeded his own education, a detail that upset him throughout his life. Indeed, he taught himself many of the things he had not been able to learn in a more formal setting. His aunt married a successful businessman, who often helped Milanés and his family financially and forms a central part of Estorino's play. It is the daughter of this aunt and uncle, Isabel—Milanés' cousin, with whom Milanés falls hopelessly in love in both the play and real life.

Milanés lived during an important historical moment that was marked by calls for independence and abolition. He traveled to Havana in 1832, looking for work but returned two years later to Matanzas. Here, he formed an important relationship with Domingo del Monte, an influential intellectual of the Cuban Romantic moment, that helped shape much of his literary work and career.[9] In 1843 Milanés reportedly fell in love with his cousin Isa. The potential union was strongly objected to by her parents, considering his lack of means and professional promise; this rejection is said to be what instigated his loss of reason in the same year. Milanés recovered briefly, but re-entered a state of delirium in 1851, in which he remained until his death in 1863 in Matanzas. While this madness virtually ended Milanés' public life, it is in many ways a source of inspiration for Estorino and others of his generation.

José Jacinto Milanés is a Romantic playwright and poet whose poems do not always adhere to strong rhetorical principles of rhyme and rhythm. His work was more focused on poetic freedom than on holding fast to lyrical norms. His lyrical work has been divided into two periods; the second, beginning in 1843, was marked by a more political focus and has generally been considered by many critics to be less worthy than his earlier poems. Ideas of these poems included, among others, socio-economic factors (access to education, distribution of wealth), abolition, and Cuba's independence. Despite critical reservations, many of these poems are laudable examples for both their themes and their lyricism. Furthermore, there is a distinct national connection in Milanés' work that in retrospect highlights the fact that he never left the island to live in exile despite his writings against slavery and for independence and the high number of other contemporary literary figures who left or were exiled under similar circumstances.

While José Jacinto Milanés is most well-known for his poetry, he was also a successful playwright. His play *El conde Alarcos* is without a doubt his most successful contribution to Cuban letters. The play, received with much critical and public praise when it debuted in 1838, was written in verse reminiscent of Golden Age drama. It takes place in Paris in the thirteenth century and the plot had had two other dramatizations in the seventeenth century, one by Lope de Vega and one by Mira de Mescua. Thus, Milanés appropriates a story that has had multiple manifestations, creating an intertextuality within the play that lives on with Estorino's appropriation of it.[10]

Estorino's fascination and use of the figure of José Jacinto Milanés characterizes much of the feeling on Milanés in at least one of the literary circles of Havana during the 1960s. As Antón Arrufat reveals in his *Virgilio Piñera: entre él y yo*, Milanés was a popular figure in many of the literary discussions in Havana at the time, particularly with Virgilio Piñera who was said to be writing something on him. Arrufat recounts how, in 1974, Estorino invited a group to a reading of his most recent play, one about Milanés. It took place at Estorino's home in Vedado, Havana with Arrufat, Piñera, Estorino, José Triana and Olga Andreu. Arrufat documents the intense interest that the figure of Milanés inspired in the group, especially in Estorino and Piñera. This parallel interest in the nineteenth-century poet suggests a manifestation in the current climate that precipitated a return to this melancholic literary figure.[11]

Given that Milanés' *El conde Alarcos* is one of the earliest plays written in Spanish in Cuba, it is only natural that Estorino focuses on him at a time of new beginnings. Vivian Martínez Tabares points out in her introductory essay to *La dolorosa historia* in the anthology *Teatro cubano contemporáneo*, how both playwrights had a deep faith in theater's ability to paint the heart of its people and their concerns in order to engage with the audience.[12] Given the importance that both authors placed on theater and their own importance to theater in Cuba, it is natural that their paths should cross literarily. Milanés' life, which, as Jorge Febles notes, has been seen to be archetypal example of a Romantic poet, offers rich material for Estorino to explore.[13]

Abelardo Estorino's *La dolorosa historia* presents the opportunity to reclaim from the colonial past the work of a forgotten poet, who abhorred colonization and slavery. Estorino returns to the past to reintroduce a Cuban literary hero who was the victim of and fought against many of the ills that the Revolution rejected. In this way, he reshapes and creates a new interpretation of Milanés that will enlarge the spectator's world. As Jorge Febles points out, Estorino demystifies Milanés.[14] At the same time, Estorino uses Milanés' biography and image to put into question 'official' stories and histories both in the past and in the present. His play encourages a critical examination of its material and that to which it alludes by undoing and rediscovering a rich past that, like an onion, reveals many layers and facets.

As stated earlier, *La dolorosa historia* did not premiere until 1985, staged then by the theater group *Teatro Irrumpe* directed by Roberto Blanco. However, there were rehearsals of the play in the seventies under the direction of Vicente Revuelta, a well-known member of the Havana theater community. According to Estorino, in an interview with me in May of 2007, Revuelta began rehearsals in another space because the theater they were using, the Hubert de Blanck, was under reparations. When the time came to move back into the Hubert de Blanck, Revuelta delayed the conversion of the play to this space and the theater needed to move on to premiere some other work.[15] There are many possible reasons for this omission, not least of which is the possibility of censorship, be it official or unofficial. As was discussed in the chapter on Virgilio Piñera, censorship was an integral part of the 1970s in Cuba and was employed through various channels.

La dolorosa historia del amor secreto de don José Jacinto Milanés is an ambitious play in both length and characters.[16] Divided into seven scenes, the first opens with José Jacinto Milanés' death and funeral procession and also introduces El Mendigo as one of the central characters of the play. The Mendigo follows Milanés, a Dante-like figure who watches his own life in the play, accompanying him along the memories of his life. In the "Prólogo [Prologue]," Milanés' coffin is brought onstage and he walks out of it. Milanés thinks dejectedly about his only literary success in his own lifetime: the premiere of his play *El conde Alarcos* in the Tacón theater in Havana and the Mendigo states that, from now on, this is what his time will consist of: "recordar y repetir, nada más [remember and repeat, nothing more]."[17] These two verbs describe Milanés' new existence, subject to what others are thinking, unable to initiate any new actions.

The play continues through key moments of Milanés' life, portraying him in Matanzas, Havana, and again in Matanzas. Scattered throughout the work are references to the present moment of Milanés' funeral, creating a dual time period of past and present and alluding to Estorino's own parallels between Milanés' time and his own. This emphasis on death and the past was of singular importance for Estorino in that it conveyed the general tone of the play, as he revealed in an interview with me in May of 2007: "es una obra que yo tenía planteada de personajes que están muertos, que reviven. Todo tendría que estar lleno de cenizas, de telaraña [it is a play that I thought of as with characters that are dead, that come to life. Everything should be full of ashes, of cobwebs]."[18] *La dolorosa historia* is a play rooted in the funereal and death; however, using this confusion of time, it connects to the contemporary moment when Estorino was writing through allusions to topics and issues central to the 1970s in Cuba.

"Delirio," the final scene of *La dolorosa historia*, explores Milanés' loss of reason and the uprising of the Conspiracy of the Ladder. The scene opens with a conversation between Milanés and Plácido, another nineteenth-century poet, where the two men compare their career and work. Though free, Plácido was subjected to many of the same prejudices against slaves, having African blood, a circumstance that limited his education and his career possibilities, but did not stop him from gaining renown as a poet.[19] This turns to the action of the slave revolt and the reader-spectator witnesses the *Fiscal* (public prosecutor) torturing a slave for information. The final name uttered by the tortured slave is that of Plácido and in the next moment Plácido is being questioned by the *Fiscal* and then is killed. In an effort to understand what slavery has done, Milanés subjects himself to be tortured and questioned by one of the slaves and then dies on stage. The Mendigo appears again and reminds both Milanés and the spectator that Milanés and his poetry live on in the memory of those who loved him and in his work and the play ends with Federico Milanés, José Jacinto's brother, reciting one of his brother's poems.

Abelardo Estorino's title alludes to many of the concrete references and lingering fears that characterized Havana in the 1970s. *La dolorosa historia del amor secreto de don José Jacinto Milanés* [*The painful history of the secret love of José Jacinto Milanés*], the full title of the play, invokes the idea of furtiveness and the pain that is provoked by this secret and its effects. This title, which, much like the play itself is very long, hides the key to the dramatic work. At face value, this seems to allude to Milanés unrequited romantic love for his cousin, Isabel Ximeno. After all, this is the unfulfilled relationship that is said to have precipitated his loss of reason. However, this 'amor secreto' also refers to Milanés' love for Cuba, a devotion which Estorino himself shared. Milanés never left Cuba despite having written much against colonization and slavery, two strongholds of the Cuban colonial government in the nineteenth century. Many other contemporary intellectuals who voiced similar opinions were exiled at some point from Cuba, forced to pursue their careers elsewhere due to harassment or in order to find more amenable conditions abroad. Milanés, however, remained committed to writing and living in Cuba, as he states in the play: "Y los amigos...los amigos...¡Ya ves! Si había que estar por la abolición había que estar hasta el final [And my friends...my friends...You know! If you were committed to abolition you had to be until the end]" (Delirio 119). Similarly, Estorino has also stayed on the island despite the exile of many other intellectuals from the same time period. Estorino's choice of a title, then, speaks to the deep dedication and commitment to the concept of Cuba that he and Milanés both share, despite political and economic hardship. However, this is a commitment that can also be seen to contribute to Milanés' reported insanity—a condition to which Estorino is calling attention in his parallel between two particularly complex historical moments.

The choice of the word "secreto" in the title alludes to the secrecy that marked the Revolution's response to intellectual dissent. Censorship can be seen as a dark and secretive process where some things are allowed to pass and others are not, as can be seen in the examples outlined earlier of José Lezama Lima and Heberto Padilla. Estorino's play is allowed to be rehearsed and appears to have passed with approval the censor's eyes. However, it never receives a proper premiere and remains hidden until the mid 1980s only seen by a select few. In this way, the "secreto" points to the very process of artistic production in Cuba at the time, a process marked by questions without answers and secret injustices.

La dolorosa historia begins with extensive stage directions that set the opening scene. They offer an important view of the direction of the work by discussing the scene changes, or lack thereof, for the entire play. Estorino directs that the play should begin with an empty stage and gradually fill up with objects that serve not only in their most obvious function as props, but also to create smaller stages on which the actors will unveil the play's story. These directions are important to understand both the play and the course which Estorino intends to follow. First, Estorino outlines the arrangement of the physical stage:

> *Al comienzo de la obra el escenario estará completamente vacío. Los muebles y la utilería serán traídos a escena siempre por negros. Una vez que se coloque algún objeto, éste debe permanecer en escena el resto de la obra, de modo que el escenario se llenará de muebles, útiles de trabajo de los esclavos, objetos de adorno, y se crearán caminos, espacios donde actuar y sentarse, aunque no sean para estas funciones, y tomará el aspecto de un lugar que ha permanecido cerrado mucho tiempo, donde nadie ha entrado. Todo debe parecer como cubierto de polvo y telarañas*

> [At the beginning of the play the stage will be completely empty. The furniture and the props will be brought onto the stage always by black people. Once an object is placed, it should remain on stage the rest of the play, so that the stage will fill up with furniture, tools for the slaves, objects of adornment. And this will create paths, spaces in which to act or to sit, although they are not for these things. And it will take on the aspect of a place that has been closed for a long time, where no one has entered. Everything should be as if covered in dust and cobwebs] (25).

At the beginning, the entire stage is empty and only fills up as the story progresses, suggesting that in the beginning there is a clean slate and, that with time, the space will come to take on meaning. Estorino's directions construct a scene that is created during the course of the play by the actual characters (the black characters, or the slaves—an essential detail, this being a play that takes place in the nineteenth century), thus eliminating the use of stagehands at scene changes. In this way, everyone that enters the stage is integral to the actual story. It is important also that the directions state that the objects will be brought on to the stage "siempre por negros [always by black people]," in that from the very beginning the reader-spectator (or director) must maintain the division of labor that the time period determines; thus, slavery is visually an important topic to the production, mirroring the actual historical moment and placing the issue of slavery at the heart of the debate.

The employment of objects in *La dolorosa historia* is central to the play in that they are used in part to create the space on the stage itself. The black slaves bring on various objects that begin to fill the empty space with which the play begins. Thus, the stage becomes more and more cluttered and disorderly as the play goes on. This is a mirror image of what is happening to Milanés himself throughout the course of the play: as the play and his life advance, his mind becomes more and more disordered, culminating in the final scene, "Delirio" which portrays the Conspiracy of the Ladder as well as his own delirious end. Estorino advises that the scenery should take on the aspect of "un lugar que ha permanecido cerrado durante mucho tiempo, donde nadie ha entrado [a place that has been closed for a long time, where no one has entered]." This forgotten aspect of the physical stage represents the metaphorical space of Milanés' own literary biography as a man who was in many ways left behind while others gained more national and international recognition (a fact due both to his own loss of reason and his desire to stay in Cuba when others went into exile). Simultaneously, this impression of 'stuffiness' that Estorino specifies refers to the temporal distance between Estorino's own historical moment and the play's. Thus, the stage for *La dolorsa historia* mirrors the deterioration that is evident in both the characters and the argument of the play.

Matías Montes-Huidobro, in "El discurso teatral histórico-poético de Abelardo Estorino: entre el compromiso y la subversión," recognizes the importance of Estorino's empty stage for the play's larger argument. Montes-Huidobro observes that this emptiness is that of the "vacío histórico [historical emptiness]" in which Estorino is placing Milanés and his biography. Estorino's aim is to re-form the life and work of Milanés before the reader-spectator's eyes.[20] Montes-Huidobro maintains that Estorino creates a historical vacuum in which he paints Milanés and his biography with the purpose of creating the nineteenth century poet as a character in the play. In this way, Estorino's work promotes a vision that creates Milanés as a character rather than as a historical figure. While I agree that Estorino is creating an empty space in which his character, José Jacinto Milanés, will interact, it is likewise important to recognize the role that history plays within this work. It is my opinion that Estorino is playing with alternate historical visions in order to provoke questions about these accepted versions. In this way, *La dolorosa historia* is more complicated than the *creation* of a character and space given that the character and the space in which he moves must come together to engage the spectators on a questioning of canonization.

Estorino's heavy authorial hand continues in the stage directions quoted above (and those quoted below) by defining precisely the movements and the presentation of the characters. These characters, much like the stage setting itself, suggest an earlier time that has long been forgotten and has likewise become withered and aged.[21] These characteristics are reinforced in both the dress and the presentation of the characters:

> *Los personajes deben recordar objetos de museo, figuras de cera en vitrinas empolvadas o momias envueltas en sudarios. Pueden estar vestidos con trajes de la época, pero en ningún momento darán la idea de riqueza o brillo, sino de algo que está desintegrándose. Las ropas estarán amarillentas, manchadas, rotas, (no por el uso sino porque han estado guardadas mucho tiempo). Los personajes estarán maquillados muy pálidos, para lograr cierto romántico aspecto fantasmal*
>
> [The characters should remind one of objects in the museum, wax figures in dusty windows or mummies wrapped in shrouds. They can be dressed in costumes from the time, but they should not at any moment give off the idea of money or brilliance, but of something that is disintegrating. The clothing should be yellowed, stained, ripped (not by use but because it has been stored for a long time). The characters will wear very pale make up, to capture a certain romantic, ghostly aspect] (25).

The characters' make-up and costume mirror the use of props: they all create a scene that marks the progressive deterioration evident in both Milanés' mental state and in his family's social standing. Estorino states that the characters should resemble deteriorated museum objects, emphasizing the distance between the subject of the play and its production and the corrosion that has come to pass through time. This quality of aged-ness and death is emphasized repeatedly here. Similarly, the characters are specifically made up to be "muy pálidos [very pale]" in order to evoke a "cierto romántico aspecto fantasmal [certain romantic, ghostly aspect]." This quality in the characters combines, to a certain degree, a Romantic idealization for a pale complexion with a ghostly quality that evokes a long-gone past that has ultimately failed in its utopian hopes. The characters reflect a historical trend while simultaneously reminding the spectators that this is a past in deterioration. Furthermore, the historical vantage point of the spectator—that of knowing that abolition and independence would come slowly and would not be the all-encompassing solution that was sought—lends the scene a mythical and spectacular air. Hidden within this disappointment in what the future will bring, the reader-spectator sees a connection between Milanés and the play and Estorino's own time. Just as independence and abolition did not eliminate Cuba's dependence on an outside government or end racism, the Revolution and the socialist ideas it brought did not offer a quick solution to the centuries of inequality or poor distribution of resources. *La dolorosa historia* is reminding its public of the difficulties of change while also pointing out the gaps between promises and realities.

The following stage directions, which open the "Prólogo [Prologue]," do even more to illustrate the scene that has been set above. It is at this moment that the reader-spectator realizes that Milanés has already died and is witnessing his own funeral. Both Milanés and the reader-spectator discover that the play will revisit certain memories of his past as others remember them. This scene opens as the funeral procession enters the stage:

> *Escenario vacío, penumbra, campanadas de duelo. Desde el fondo del escenario avanza el cortejo de un entierro; los personajes musitan o cantan un poema de Milanés; traen un libro con sus obras; al llegar al frente se abren en dos filas y van hacia los lados. Al fondo queda el ataúd, vertical. El Mendigo se acerca y lo abre; Milanés descruza las manos que tiene sobre el pecho. El Mendigo lo toma por una mano y lo hace avanzar algunos pasos*
>
> [Empty stage, semi-darkness, mourning bells. A funeral procession advances from the back of the stage; the characters whisper or sing a poem by Milanés; they carry a book of his works; when they arrive at the front they break into two rows and move towards the sides. At the back there is the coffin, placed vertically. The Beggar approaches it and opens it; Milanés uncrosses his hands from his chest. The Beggar takes his hand and makes him advance a few steps] (Prólogo 26).

The scene opens on a mournful note, suggesting that the play will not be a simple exaltation of Milanés' life but will question its circumstances. Nevertheless, there is a space to honor Milanés' memory and poetry given the somber manner with which the processants enter the stage. Throughout the play, Estorino quotes sections of works or entire poems that violently cut into his own text. This forces the reader-spectator to remember another time period and context and bridge the gap between the two. Furthermore, it lends depth to *La dolorosa historia* itself and allows Estorino to say much more through the use of intertextuality than would be otherwise possible.

Montes-Huidobro returns to Estorino's stage directions at the beginning of the play's prologue to underline the importance of the role of temporality in *La dolorosa historia*. Similar to how he identified a move towards a historical emptiness in the stage directions that open the play, Montes-Huidobro maintains that the play puts forth a "vacío temporal [temporal emptiness]" that breaks any linear identification with time despite being a biography. This highlights Estorino's break with the traditional idea of linearity in order to place the characters in a space beyond time that will allow the play to function more freely in the past, present and future.[22] Montes-Huidobro is right in identifying the centrality that time and its fluidity occupy in the play. There is much movement on the part of the characters through memories that take place at different moments, shifting the focus from chronology to fluidity. Similarly, there is much confusion between the period in which the action takes place and that in which it is written: that is, José Jacinto Milanés' time period versus that of Abelardo Estorino. There are many moments when the two playwrights seem to identify and connect across the century that separates them. In this way, there is a "vacío temporal" that attempts to function across the past, present, and future and its objective is to question and to provoke through the use of these historical issues from the mid-nineteenth century.

The opening scene, as observed in the quote above, uses a coffin to illustrate Milanés' death, thus immediately signaling the importance that death will play in *La dolorosa historia*. This lends a more somber air to the scene, one that is compounded by the fact that Milanés steps out of the coffin in order to observe his own funeral. Milanés travels throughout the play as a sort of visitor to his own life and the memories that make up his life. Despite the seemingly faultless transition that the poet had undergone to death, it is not completely peaceful, and he rails against being taken from those he loves and being forced to revisit painful memories. Milanés cannot leave his current state to return to those who took care of him but becomes the victim of others' memories of him and his life, narrating their actions and his own emotions:

> *Milanés*: Vete. Carlota me pone compresas frías, compresas frías. Tengo fiebre, me ahogo, vete. (*El Mendigo va hacia él. Milanés huye.*) Carlota, despiértame, ábreme los ojos, ábreme los ojos, los ojos, Carlota. (*Se cubre los ojos, el Mendigo se acerca.*) Vete, no quiero verte.
>
> *El Mendigo se aleja. Milanés se queda en el centro con los puños sobre los ojos. Un actor del cortejo se acerca con una estaca en cuyo extremo está clavada la cabeza de un negro. Da vueltas alrededor de Milanés. Los otros personajes del cortejo restallan látigos. El actor clava la estaca junto a él y cesa el sonido de*

los látigos. Silencio. Milanés abre los ojos y al ver la cabeza grita: Sálvame. El Mendigo se lleva la estaca y vuelve junto a él.

Mendigo: Ya, ya pasó.

[*Milanés*: Go away. Carlota puts cold compresses on me, cold compresses. I have a fever, I'm suffocating, go away. (*The Beggar goes towards him. Milanés flees.*) Carlota, wake me up, open my eyes, open my eyes, my eyes, Carlota. (*He covers his eyes, the Beggar moves towards him.*) Go away, I don't want to see you.

The Beggar moves away. Milanés remains in the center with his fists over his eyes. An actor from the procession moves towards him with a stake on which a black man's head is nailed to one end. He circles around Milanés. The other characters of the procession crack whips. The actor nails the stake close to him and the sound of the whips stops. Silence. Milanés opens his eyes and, when he sees the head, screams: Save me. *The Beggar takes the stake away and goes back to him.*

Beggar: It's over, it's over] (Prólogo 30).

Milanés longs for Carlota, the sister who cared for him during the long years of his illness. Simultaneously, he rejects the Mendigo, a character that he created in a poem entitled "El mendigo [The beggar]" hoping to expel the thoughts embodied in the image out of his mind. Nevertheless, instead of being liberated from these torments, Milanés is haunted by visions of torture, embodied in a whip and an executed slave's head—images which become real at the end of the play. As seen here, the violence of *La dolorosa historia* is not just a physical one of torture, but encompasses the inner demons that helped to drive Milanés into insanity. Violence here is both physical (the dead slave's head) and psychological (the haunting that Milanés experiences), each one more horrific than the other. Indeed, the very premise of the play is a violent experience where Milanés is forced to relive his own excruciating moments while accompanied by a character he originally created in order to free himself of the shameful memory that this figure first inspired.

The Mendigo is a central character in *La dolorosa historia* who accompanies Milanés through his journey of memories, clearly referring to the poem "El mendigo" that Milanés wrote in 1837.[23] Here, the poet discusses an encounter with a beggar at the entrance to a dance. In the sixteen stanzas of the poem, he contrasts the opulence of the ball ("La casa de baile muy bella lucía: // todo era cortinas y luces y espejos, // y damas vistosas entrando a porfía // y música dulce sonando a lo lejos: [The dance hall looked very beautiful: // everything was curtains and lights and mirrors, // and showy ladies entering in an obstinate manner // and sweet music that could be heard far off:]") with the presence of a beggar asking for charity as the lavishly dressed, young men enter the hall ("Alegres mancebos entraban conmigo // cuando al ir entrando, tendida a nosotros // la pálida mano de anciano mendigo // pidiónos limosna, negada por otros. [Happy young men entered with me // when upon entering, stretched out in front of us // the pale hand of an old beggar // asked us for charity, denied by the others.]")[24] Among Milanés' commentary on the contrast in socioeconomic class, there is criticism directed at the poet himself and against the society in general in which Milanés moved given that he answers "como todos [like everyone]": "Hecho ya al idioma cruel del agravio // me mira el anciano y ante mí se pone; // mas yo, vergonzoso, con trémulo labio, // le di como todos mi estéril *perdone*. [Already made cruel by the offense // the old man looks at me and places himself in front of me; // but I, ashamed, with a tremulous lip, // gave him, like everyone, my sterile *excuse me*.]" This brief encounter with the beggar stays with the poet throughout the dance and shames him, not just because of the ostentatious show that contrasts with the beggar's lack ("Y ostentaban todas, que era fácil verlas, // sus perlas, sus trajes, como hace una actriz // sin ver que brillaban sus nítidas perlas // cual lágrimas tristes de un hombre infeliz [And all of them showed off, it was easy to see them, //their pearls, their dresses, as an actress does // without seeing that their shining pearls shimmer // the sad tears of an unhappy man]"), but also because of Milanés' fears that his own lack of means is revealed in his rebuff of the beggar: "Si acaso pasaba riendo un amigo, // creí escucharle que hablaba de mí. // "Ved: ése no tuvo qué darle al mendigo, // y viene a reírse y a danzar aquí" [If by chance a laughing friend passed by, // I believed I heard him speaking of me. // 'See: that one didn't have anything to give the beggar, // and he comes to laugh and dance here']." The figure of the Mendigo haunts Milanés throughout the night, which he tries to escape by writing this poem. In a cruel move that eliminates the escape for which Milanés strove by writing the poem, Estorino chooses to use this image to accompany the poet on his journey, underlining the reciprocal relationship that exists between the creator and its object by turning this on its head.

"El mendigo" is an important poem in Milanés' lyrical production in that it shows both his poetic ability and his growing social conscience, an interesting combination that Estorino chooses to highlight in his twentieth-century reproduction of the nineteenth-century poet and playwright. The Mendigo maintains the role of a guide through Milanés' emotions and conscience, but here he deepens this responsibility to become a companion to the poet along the difficult path of memories:

> *Milanés*: [...] (*Mira al Mendigo y reconoce al personaje de un poema.*)
> ¿Quién eres?
> *Mendigo*: ¿Ya me recordaste?

Milanés: Siempre me dio miedo.
Mendigo: Entonces no debías haber escrito el poema en que aparezco.
Milanés: Quería liberarme del espanto y ahora estás aquí.
[...]
Milanés: ¿Por qué estás conmigo?
Mendigo: Alguien piensa que debo acompañarte.
[*Milanés*: [...] (*He looks at the Beggar and recognizes the character of a poem.*)
Who are you?
Beggar: Now you remember me?
Milanés: You always made me afraid.
Beggar: Then you shouldn't have written a poem in which I appear.
Milanés: I wanted to free myself of the fear and now you're here.
[...]
Milanés: Why are you with me?
Beggar: Someone thinks I should accompany you] (Prólogo 29–30).

In this scene, Milanés remembers the fact that he was haunted by the figure of the Mendigo and was compelled to write a poem about him in order to liberate himself from the images. While Milanés remains a sensitive figure haunted by the compelling images that surround him, the Mendigo takes on a more active and powerful role in Estorino's *La dolorosa historia* in that he becomes the poet's guide through the riveting and violent memories that comprise the play. In fact, it is interesting to incorporate a character that the Milanés of the play admits he created in order to forget. Estorino's inclusion and empowerment of the Mendigo in this play about José Jacinto Milanés' life suggests that Estorino's objective is not to retell what is already known, but to delve deeper in order to reveal the complexities of poverty, social marginalization, and mental deterioration that make up this life. Milanés, then, becomes the means to arrive at other ideas and issues, such as the Revolution and its promises, the writing and use of history, and current intellectual practices.

Despite the intimacy that the Mendigo can claim to the poet's life, Milanés himself questions his presence at this crucial moment. The Mendigo's response, "Alguien piensa que debo acompañarte [Someone thinks I should accompany you]," reveals the machinations that are at work behind the scenes in the play's presentation of Milanés' biography. For Montes-Huidobro, this "alguien" is Estorino himself, the engineer of the play and of the play's characters.[25] Montes-Huidobro identifies correctly that Estorino is creating his own fiction of Milanés' life from the details that he has been able to gather. Thus, to carry this one step further, Milanés steps out of the biography of the nineteenth-century poet to become a character that Estorino is able to use to explore Cuba's nineteenth century and parallel it with his own historical moment. This purpose is the fundamental question for the play and my reading of it. Estorino is the one to connect once again Milanés with the Mendigo. In what can be interpreted as a violent gesture, Estorino brings the characters together again in order to highlight both this contrast in economic distribution (one that is not limited to the nineteenth century) and the two characters themselves—one the creator, one the created. In *La dolorosa historia*, this relationship transforms itself into something new, where both are creations of something else and thus gain autonomy.

Death haunts much of *La dolorosa historia*. Indeed, the very play begins with the death of Milanés, thus reversing the normal order of a biography and almost personifying death as a character that will reappear in the play. Throughout the seven sections, it remains a principal topic of the play and intermingles with memory. Estorino underlines how death formed a significant part of Milanés' childhood in that many of his brothers and sisters died before they could reach adulthood. As seen in the conversation quoted below between Milanés and his mother, death helped to mark and form who this nineteenth century poet would become:

> *Doña Rita*: ¡Qué sabes tú lo que es parir quince hijos! Veintiún años estuve así. (*Se toca el vientre.*) Uno tras otro, uno tras otro.
> *Milanés*: Ocho murieron.
> *Doña Rita*: Ay, Pepe, ¿eso qué importa ahora? Ya todos somos recuerdos...recuerdos...
>
> [*Doña Rita*: What do you know about giving birth to fifteen children! I was like this for twenty-one years. (*She touches her belly.*) One after another after another.
> *Milanés*: Eight died.
> *Doña Rita*: Oh, Pepe, what does that matter now? We're all just memories...memories...] (La familia 33).

Similar to the opening of the play when Milanés' funeral procession enters the stage, the phantom of death enters the play again, but here this scene explains the poet's fascination with death and why it forms a prominent part of *La dolorosa historia*. Doña Rita's response to Milanés' observation signals the importance death has had for Milanés in contrast with its significance for herself and others, for example. At the same time, her words reflect the role of memory in creating both a life and this particular play.

Death does not just haunt the characters in *La dolorosa historia* but seems to invade the very location of the play. Milanés (and Estorino, for that matter) was born in Matanzas, a city and province that connote death in its very name and history. In the scene quoted below, Milanés is returning to Matanzas from Havana. The scene opens with the Mendigo, Zequeira, Josefa la Endemoniada [Josefa the Possessed], and el Sereno [the

Night Watchman], discussing Milanés' imminent return and the founding of his birthplace through violent acts:[26]

> *Sereno*: Y autorizó que treinta familias canarias ocuparan las tierras al borde de la bahía donde los indios perpetraron la matanza que dio nombre a la región.
> *Josefa*: Y Matanzas era la bahía.
> *Mendigo*: Y Matanzas el río que después fue San Juan.
> *Zequiera*: Y Matanzas las tierras frente a la bahía.
> *Sereno*: Y ahora Matanzas es tu ciudad, como lo fue de tus ascendientes, castellanos de San Severino y alcaldes de la Santa Hermandad.
> *Josefa*: La Matanzas de tu antepasado José Ignacio Rodríguez de la Barrera, cura de la iglesia de San Carlos, enviado especial del Santo Oficio, que vino a la ciudad buscando herejes.
> *Zequeira*: Tus antepasados eran los más puros, los que podían descubrir el demonio de los otros. De esa cepa vienes.
> *Sereno*: No niegues la tradición. Ven y establece la pureza en la ciudad.
> *Josefa*: Yo profetizo: en esta ciudad se cometerá la mayor matanza de negros en nuestra historia. Los perseguirán como fieras, los atarán a una escalera y los azotarán hasta desangrarlos. Ven, no te pierdas ese espectáculo, aprende a hacer historia.
>
> [*Sereno*: And he authorized thirty families from the Canaries to occupy the lands on the edge of the bay where the Indians perpetrated the massacre that gave the region its name.
> *Josefa*: And Matanzas was the bay.
> *Beggar*: And Matanzas the river that was later San Juan.
> *Zequiera*: And Matanzas the lands across from the bay.
> *Sereno*: And now Matanzas is your city, just like your ancestors, Castilians from San Severino and mayors of the Santa Hermandad.
> *Josefa*: The Matanzas of your ancestor José Ignacio Rodríguez de la Barrera, priest at the church of San Carlos, especially sent by the Holy Inquisition, that came to the city looking for heretics.
> *Zequeira*: Your ancestors were the most pure, the ones that could discover the devil in others. Those are your roots.
> *Sereno*: Don't deny tradition. Come and establish peace in the city.
> *Josefa*: I profetize: in this city, the biggest massacre of blacks in our history will be committed. They'll pursue them like demons, they'll tie them to ladders and whip them until they bleed. Come on, don't miss that spectacle, learn to make history] (Matanzas 54–55).

In this scene, the four characters focus their attention on the city of Matanzas, Milanés' home. Their exchange concentrates first on locating Matanzas and then turns to the violent events that have formed the city. Their conversation recalls the distinguished past of Milanés' family in the province of Matanzas and parallels, in this way, Milanés' own personal history of loss of status. Thus, the characters inscribe him within a tradition of violence that cannot help but leave its mark on the poet and playwright.

The characters in this scene are important, given that they turn attention to their own role within this biographical work. In the beginning of *La dolorosa historia*, Estorino classifies his *Dramatis Personae* into nine different categories depending on their relationship to Milanés. These four characters are classified within "La imaginación de Milanés [Milanés' imagination]." This classification marks them as different from others who fall into groups such as "La familia de Milanés [Milanés' family]," "Los amigos [Friends]," or "Los negros [The Blacks]," despite the fact that Zequeira, for example, did exist. Nonetheless, Estorino puts them all within the category "La Imaginación de Milanés," an important point given that this poet is dead and can only move through Milanés' memories in his imagination. This is an essential detail that, when joined with the fact that Zequeira was also a poet who suffered a loss of reason that accompanied him to his death, sheds light on the groupings. Both Milanés and Zequeira, along with Josefa la Endemoniada, are marginalized figures who, because of their madness, do not function as others in the play do. The Mendigo, in turn, is the product of Milanés, a man without reason, and thus belongs with the marginalized.

Taking full advantage of the literary past, Estorino uses fragments of Milanés' literary work at moments throughout *La dolorosa historia* to connect a certain life memory with the nineteenth-century poet's lyrical work. Quotation is an important strategy that here amounts to a violent gesture that blurs literary genres. The first fragment of a poem by Milanés appears in the "Prólogo" and is provoked by the smell of flowers:[27]

> *Mendigo*: [...] ¿Te llega el olor de flores?
> *Milanés*: Azucenas.
> *Mendigo*: Sí, había muchas. Y dalias, dalias enormes, rosas, madreselvas, todas blancas. Flores blancas llenaban la casa.
> *Milanés*: Cuando mi hermano menor
> huyó tronchado en su flor
> de este universo ilusorio,
> le mandó mi padre ornar
> de flores, y rodear
> con los cirios del velorio
> *Mendigo*: ¿Quién estará recordando esos versos?
> *Milanés*: Yo los recuerdo.
> *Mendigo*: (*Suelta una carcajada*.) No recuerdas ni versos, ni flores, ni campanas, ni sollozos. Nada.
>
> [*Beggar*: [...] Do you smell flowers?

Milanés: Lilies.
Mendigo: Yes, there were many. And dahlias, enormous dahlias, roses, honeysuckle, all white. White flowers filled the house.
Milanés: When my younger brother
fled cut short in his flower of life
from this illusory universe,
my father sent him to decorate
with flowers and to surround
with candles the wake
Mendigo: Who could be remembering those lines?
Milanés: I remember them.
Mendigo: (*He laughs*.) You don't remember lines, flowers, bells, or sobs. Nothing] (Prólogo 26–27).

As the spectator-reader can observe from this scene, the poem is seemingly inspired by the smell of flowers, but the Mendigo reveals that Milanés is reciting his poem that has been inspired by what *someone else* is remembering. Milanés' memories are no longer controlled by himself but by outside forces. The role of outside influences is a common element of all artistic works. Milanés' literary production, as we have seen, was often aided by outside forces (prompted by the presence of Domingo del Monte, inspired by socio-economic conditions or the smell of flowers, among other things) as was the work of many writers in the moment Estorino was writing.

The inclusion here of a poem within the lines of a play is not completely unusual, given that the two genres often intermingle. Indeed, plays were often written in verse, thus making even more natural the connection. In *La dolorosa historia*, however, Estorino is connecting to Milanés' poetry through the thoughts of someone else, violently robbing the nineteenth-century poet of control over his own memories. The fact that this poem comes to Milanés' mind because another is thinking of it violently connects Milanés' thoughts with those of someone else while also mixing genres to create a fusion that questions borders between one person or thing and another.

The appropriation of Milanés' poetry (both the actual verses and the characters—such as the Mendigo) blurs the lines between the authors (Estorino and Milanés) and between the genres (drama and poetry). For Montes-Huidobro, the use of Milanés' poetry in Estorino's play is another illustration of the intermingling and experimentation that is present in the play. The poems, for Montes-Huidobro, intertwine two spaces and time periods, thus confusing them and making them into one.[28] As Montes-Huidobro identifies, contradiction in *La dolorosa historia* is the very objective that Estorino intends to use to illustrate his meaning.[29] Estorino's innovation, for Montes-Huidobro, comes from the blending of borders in his play. *La dolorosa historia* tries to undo genres and groups in order to move beyond classifications. Estorino uses another period and playwright to force the spectator to look again at the present time and situation. In this way, a dialogue is initiated with the public that attempts to revisit history to understand more fully both the events and their consequences.

In the final section of the above scene, the Mendigo reveals that Milanés' thoughts and memories now materialize from the memories of those who are still alive and remember him. Milanés cannot remember his life on his own but is prompted by how others are thinking of him. This concept, a difficult one to understand at first for both the spectator and Milanés, is made known progressively throughout this scene by the Mendigo and brings both loving and painful memories with it as Milanés discovers that, though he is gone, he will be remembered:

> Tienes que acostumbrarte. Ellos seguirán recordándote: Carlota, Federico, harán un culto a tu memoria; publicarán tus versos una y otra vez; contarán anécdotas, recordarán tu niñez, la escuela, los primeros versos, después el éxito...
>
> [You have to get used to it. They'll continue remembering you: Carlota, Federico, they'll make a cult to your memory; they'll publish your poems again and again; they'll tell anecdotes, they'll remember your childhood, school, your first poems, afterward success...] (Prólogo 28).

Milanés' journey is determined by other people's thoughts and must bend to what they think, a frightening concept despite Milanés' last years of delirium, as can be seen when the Mendigo (Milanés' own creation) consoles him with a hug. Nevertheless, Estorino's assertion that the deceased are subject to the memory of the living is an important observation, especially in reference to Milanés. Federico Milanés, Milanés' younger brother and the only other male child to survive to adulthood, was also a poet, though he is mostly remembered today for either his poetry dedicated to his brother's memory or for the fact that he edited his brother's works.

Forgetting is a common fear for Milanés throughout *La dolorosa historia* despite (or perhaps because of) the fact that the entire play is formed of memories of Milanés' life. He worries at various points that the living will forget him, a fear that both Federico (and Carlota in another scene) tries to assuage in him:

> *Federico*: Pasará el tiempo y yo contaré que te quedabas leyendo y diré que tuviste que aprender solo porque no teníamos dinero para ir a un buen colegio.

Milanés: Pasará el tiempo y lo leerán y nos criticarán y no comprenderán todo el trabajo que nos costó vivir aquella época. Y después nos olvidarán.
Federico: Yo haré que no se olviden.
Milanés: Siempre confié en ti.
[*Federico*: Time will pass and I'll tell how you would continue reading and I'll say that you had to learn on your own because we didn't have money for you to go to a good school.
Milanés: Time will pass and they'll read it and they'll criticize us and they won't understand how hard it was to live at that time. And then they'll forget us.
Federico: I'll make sure they don't forget.
Milanés: I always trusted you] (La familia 40).

In this scene, Milanés fears that he and his work will fade into oblivion, a fate that his brother, Federico, promises not to allow to happen. Federico narrates to Milanés what in the future he will do to safeguard his brother's image from vanishing. The verb tenses used here are revealing, given that the men are using the future tense to describe what has already happened in the past when they utter the words—they are speaking with the future tense after Milanés' death, about actions that Federico has already done. When Federico says, "Yo haré que no se olviden [I'll make sure they don't forget]," he is describing his life's work after his brother's delirium and death.[30] His words to Milanés remember what he *will* do. In contrast, when Milanés says "Siempre confié en ti [I always trusted you]," he *is* expressing his past trust in his brother's word and devotion. This multiplicity of moments in the brothers' dialogue reveals a play on time in *La dolorosa historia* that is emblematic of the play's dislocation of time frames.

Milanés' words about not understanding the past connect also to Estorino's own time in the 1970s when many artists and intellectuals in Cuba were under extraordinary pressure to create works that would adhere to the Revolution's restrictive definitions of artistic and literary production while remaining faithful to their own artistic designs. Echoed in these words, the reader-spectator can hear Estorino's and other artists' desire to be understood by those who would read them in the future. Like Milanés, Estorino is writing at a moment of great political and social definition and turmoil, a time when outside forces may influence the words or ways that writers may describe their worlds. These outside pressures can shape the artistic works of their time in a way that demands a reading that connects these forces. With these words, both Milanés and Estorino are underlining the connection between an artist and the context in which he creates.

Even with Federico's reassurances, Milanés continues to be consumed with the idea that his legacy will be lost to oblivion. Despite Milanés' fears of being forgotten, Carlota and Federico remain true to his memory—oftentimes at great sacrifices to their own success and happiness:

> *Federico*: Nosotros no te olvidamos nunca.
> *Carlota*: Guardamos tus papeles.
> *Federico*: Imprimimos tus poemas.
> *Carlota*: Te llevamos flores al cementerio.
> *Federico*: Conservamos tu cuarto como lo tenías.
> *Carlota*: Dedicamos el resto de nuestros años a tu memoria.
> *Federico*: La casa se convirtió en altar.
> *Carlota*: Rechacé a los pretendientes.
> *Federico*. Tus poemas se hicieron populares.
> *Carlota*: Y me vestí siempre de negro.
>
> [*Federico*: We never forgot you.
> *Carlota*: We saved your papers.
> *Federico*: We printed your poems.
> *Carlota*: We took flowers to you at the cemetary.
> *Federico*: We kept your room as you had it.
> *Carlota*: We dedicated the rest of our years to your memory.
> *Federico*: The house was converted into an altar.
> *Carlota*: I refused suitors.
> *Federico*: Your poems became popular.
> *Carlota*: And I always dressed in black] (Matanzas 58).

Both siblings profess here that they never forgot their deceased brother and did all they could to ensure that, through their actions, others would not forget Milanés and his work. Their words testify to the fact that they preserved his memory against oblivion through a violent suspension of what they themselves were living in order to remember their brother.

Milanés' preoccupation with forgetting contrasts sharply with the role of his own memories in the play. Since what he remembers is controlled by those who remember him, Milanés cannot prevent re-experiencing painful memories—particularly those that Estorino brings up from his trip to Havana and his subsequent return to Matanzas.

> *Milanés*: ¡Basta! No quiero recordar más.
> *Mendigo*: Eso te tocó vivir.
> *Milanés*: Qué dolor esa ciudad perdida. Y hay otros sucesos esperando, lo sé.
>   Como la caja de Pandora, levantas la tapa y salta la sangre.
>
> [*Milanés*: Enough! I don't want to remember anymore.
> *Mendigo*: That's what you lived.
> *Milanés*: What pain in that lost city. And there are other things waiting, I know. Like Pandora's box, you lift the top and blood splatters] (Tertulia 68).

Remembering is not a particularly easy task for Milanés. His memories, not controlled by him nor tainted with the rose-colored lenses that many use to remember, become a violent monster that haunts him throughout the play and causes him to relive the very moments that are most painful. In this way, death wrenches him from the madness that allowed him to live in his own world, free of the quotidian pain that accompanies life.

Theater is another part of the play's meta-discourse. In the scene "Tertulia," the discussion turns to theater by reproducing Domingo del Monte's *tertulias* (or literary gatherings), which brought together many of the great literary minds of the Cuban nineteenth-century, del Monte, Ramón de Palma, and Cirilo Villaverde, to talk about the topics of abolition and independence.[31] There is a juxtaposing split in the action on the stage between the conversation at the *tertulia* and the physical labor of the slaves that emphasizes the central project of *La dolorosa historia*: while the men are talking, the scene focuses on a group of slaves working and the overseer watching over their work. The four literary men discuss the role of theater in society and in the Cuban context in particular. This discussion is of particular interest in the life of Milanés and the play *La dolorosa historia* as well as the contemporary context in which Abelardo Estorino is writing, given the many parallels that can be traced between the two periods:

> *Milanés*: El teatro es más difícil que la poesía.
> *Villaverde*: Sí, es cierto. Los pueblos nuevos viven más la vida del sentimiento o la poesía, que la vida del juicio o la meditación.
> *Milanés*: El drama no sólo debe pintar el exterior del hombre sino también su interior. Y entre nosotros debe expresar una deducción moral que nos saque de la impasibilidad en que vivimos.
> *Palma*: No podemos tener teatro: somos un pueblo sin historia.
> *Del Monte*: Cállese, pesimista a la moda. (*Risas*.) Discutí mucho con Heredia: hay la posibilidad de un teatro Americano, olvidándose del fatalismo griego. Huáscar, ése es un tema; Huáscar atrayéndose la cólera de su padre, las disensiones de Huáscar y Atahualpa, la sangrienta jornada de Cajamarca.
> *Villaverde*: No sé qué pensar. Hay escritores y público que no están dispuestos a escribir ni a oír hablar de otra cosa que de dinero, de negocios, y de empresas. Y si acaso de diversiones, chistosas o ridículas, cuando no escandalosas.

> [*Milanés*: Theater is harder than poetry.
> *Villaverde*: Yes, that's true. New lands live more the life of sentiment or poetry, than the life of judgement or meditation.
> *Milanés*: Drama shouldn't only portray the exterior of man but also his interior. And between us it should express a moral deduction that jolts us out of the impassivity in which we live.
> *Palma*: We can't have theater: we are a people without history.

> *Del Monte*: Be quiet, fashionable pessimist. (*Laughs*.) I talked a lot with Heredia: there is the possibility of a theater of the Americas, without Greek fatalism. Huascar, that's a topic; Huascar drawing his father's anger, the dissention of Huascar and Atahualpa, Cajamarca's bloody day.
> *Villaverde*: I don't know what to think. There are writers and a public that are not capable of writing or listening to anything other than money and business. And maybe of funny or ridiculous diversions, as long as they're not scandalous] (*Tertulia* 70–71).

These four men use the idea of theater to discuss their national context and its potential for independence. It is interesting that the subject of their discussion is exactly the product that the spectator-reader is taking in. Theater here is explored as a viable literary option in the Cuban colonial context.[32] Milanés (perhaps echoing Estorino's own thoughts) believes it to be a difficult project that should examine all sides of its characters and subject matter.

After this discussion on the purposes of theater, Milanés presents a scene from his famous play *El conde Alarcos*. In Estorino's dramatization of a scene of *El conde Alarcos*, Milanés reads the part of the count and El Español [Spaniard] plays that of El Rey [The King]. In the written text, however, Estorino designates "ALARCOS" and "REY" for only the first two exchanges and then uses "Milanés" and "EL ESPAÑOL" to mark who is speaking, effectively collapsing Milanés' dramatic identity with that of Alarcos. The written text mirrors what would be obvious for the spectator: a confusion of Milanés with Alarcos and El Español with El Rey. This detail obscures the use of Milanés' play within Estorino's play and creates a blending of the two plays and the two different sets of characters, suggesting that the master-slave relationship that Milanés portrays in his adaptation of a thirteenth-century story can see its mirror image in his own nineteenth (and, then, in Estorino's twentieth) century. On a further note, the scene that is reproduced is where the count admits to being married and the king orders Alarcos to murder his wife. The king states that the count is his slave and thus must do what he orders: "Tú eres esclavo mío. // Conde, no hay más que decir // sobre lo dicho. Ella tiene // esta noche que morir [You are my slave. // Count, there is nothing more to say // about what's been said. She must // tonight die]" (*Tertulia* 89). By highlighting this language in the scene, Estorino underlines Milanés' commitment to freedom along side his condition of marginalized figure.[33] The twentieth-century playwright is pointing out that Milanés was fully aware of the parallels between the centuries of slavery in all its forms, a connection between time periods that Estorino is also underlining in his own play. Like Milanés connecting his nineteenth century to Alarcos' thirteenth,

Estorino uses the debates on abolition and independence of the nineteenth century to focus on the ideas of Revolution and dissenting voices in his own 1970s. These confusions of time periods and characters offer a way of refocusing attention to the eternal debates on politics and social context. In this way, the reader-spectator is prepared to make connections between historical events, a connection that will culminate in the final scene of the play.

In *La dolorosa historia*, violence manifests itself in many ways; physical violence is one of the most prominent, as can be seen at the end of the play during the slave rebellion in Matanzas and its bloody repression by the government and local plantation owners in the Conspiracy of the Ladder. The brutal violence that this entails seems to infect everything surrounding the events and makes it impossible to exist outside of it. In the following scene, Pastora, Milanés' aunt, enters with a large knife, the physical evidence of the violence that surrounds them. The two frantically verbalize their own reactions to the bloodshed that marks their surroundings:

> *Pastora*: Hay que limpiar. Buscaré agua y jabón y no quedará una sola mancha.
> *Milanés*: La sangre no puede limpiarse, se adhiere a las cosas en coágulos cárdenos.
> *Pastora*: Agua, mucha agua. No quedará una sola mancha. Quiero que todo sea impoluto y reluzca.
> *Milanés*: A mis niñeces volvedme gratas,
> que ya volaron como nubes.
> (*Transición*.) Es inútil.
> *Pastora*: Me destrozaré las manos purificándolo todo.
> *Milanés*: Siempre queda un coágulo oculto. Es mejor levantar el cuchillo y... (*Se abre el cuello de la camisa, se palpa buscando un lugar.*) ¡aquí!
> [*Pastora*: We must clean. I'll find soap and water and there won't be a single spot.
> *Milanés*: Blood can't be cleaned, it adheres to things in purple clots.
> *Pastora*: Water, a lot of water. There won't be a single spot. I want everything spotless and shining.
> *Milanés*: To my childhood return me happily,
> which already flew away like clouds.
> (*Transition*.) It's useless.
> *Pastora*: I'll destroy my hands purifying all this.
> *Milanés*: There will always remain a hidden clot of blood. It's better to raise the knife and... (*He opens the neck of his shirt, he feels around looking for a place.*) here!] (El amor 102).

Here both characters express their own exasperation and incomprehension of what has happened. They speak at, rather than with, one another in disbelief and horror without seeming to hear or understand what the other is saying. Though both react strongly against the violence that has taken place, these reactions are also violent in their strength—their words and actions are extreme. Milanés and Pastora mirror one another in intensity, but not in action. Pastora's reaction is to scrub desperately the surfaces that have been soiled, destroying her own body in order to erase what has happened. Milanés, in turn, sees no way to erase the bloodshed around him but by offering himself as a sacrificial victim in exchange, an action with which he will follow through in the last scene of the play. Like René Girard stated in *Violence and Sacrifice*, despite their differences, both responses show that violence must be answered with more violence. Their violent intensity suggests that the only response to violence is more violence, creating a circularity of violence that does not allow its victims to be released from its burden.

This circularity of violence is underlined by the memories of another dramatic scene in Revolutionary Cuban theater: that of Lalo from *La noche de los asesinos* (1965) with the knife that he supposedly used to kill his parents at the end of the first act. The image of Pastora with the bloody knife recalls this pivotal scene and though the responsibility of the act is different in the two plays, the connection is identified in how the simple image of the knife poignantly recalls the three siblings and their violent efforts to break free from their parents' yoke. In *La dolorosa historia*, the knife similarly connotes ideas of intense violence and circularity, alluding both to Cuba's past and present.

The last scene "Delirio [Delirium]" is the most important one in the play. It is here that Milanés' life culminates alongside the historical events from 1844 that permanently marked the province of Matanzas. The scene opens with an encounter between Plácido and Milanés in which the two talk about their nation's position within their history and their future. Plácido attracted the negative attention of the Cuban government and elite due to his poetry and was targeted during the Conspiracy of the Ladder. He was executed publicly in 1844 in Matanzas; it is at the moment after his execution that Milanés and he discuss the political situation in "Delirio." At this point in the play, Plácido has just been executed and his head and shirt are full of his own blood, making him a grotesque physical reminder of the brutality and violence that marked the repression. Plácido points out a similarity between the two men: "Hay algo que nos iguala, mi muerte y tu delirio [There's something that equals us, my death and your delirium]" (Delirio 104). At the same time, Plácido calls attention to the differences between their lives and the privileges that Milanés has enjoyed thanks to his race and social situation, compared to Plácido's existence as an illegitimate mulatto during slavery. The two first express disbelief at the savagery in slavery and they both condemn these actions, though their words recognize the different level of guilt and complicity that the two have in the present situation:

> *Milanés*: Me asombra la gente que goza viendo cómo dos animales se destrozan.
> *Plácido*: Odio a la gente que goza atando a un negro.
> *Milanés*: Yo también.
> *Plácido*: Lo sé, por eso puedo hablar contigo. No estoy tan envilecido.
> *Milanés*: Perdóname.
> *Plácido*: Te perdoné hace tiempo.
> [*Milanés*: I'm shocked by people who enjoy watching two animals destroy one another.
> *Plácido*: I hate people that enjoy whipping a black man.
> *Milanés*: Me too.
> *Plácido*: I know, that's why I can talk to you. I'm not that debased.
> *Milanés*: Forgive me.
> *Plácido*: I forgave you a long time ago] (Delirio 105).

While Milanés' reaction denounces the thirst to see blood that he identifies in some people, Plácido reminds him of what man does to his fellow man because of race. Milanés' answer is to ask for Plácido's forgiveness, thus recognizing his complicity in slavery and racial injustice. Nevertheless, their conversation continues and turns to more personal issues. Milanés questions Plácido's decision to write what he considers inferior poetry, not being able to understand Plácido's deeper reasons and marking a fundamental difference between the two poets:

> *Milanés*: Escribí aquel poema irritado al cómo desperdiciabas tus dotes. (*Molesto*.) ¿Cómo podías escribir aquellas odas, cantar el cumpleaños de una niña tonta, ensalzar a un viejo gordo y gotoso cargado de dinero? No puedo entenderlo.
> *Plácido*: Es muy simple. Tenía ruidos en la barriga y había que llenarla, si no el estruendo cubriría la Isla. (*Tono confidente*.) Y podían acusarme de subversivo. Infidencia, es la palabra exacta.
> *Milanés*: Yo tampoco era rico.
> *Plácido*: Pero tú eras blanco.
> *Milanés*: Había que ser inflexible, no ceder ante la corrupción.
> *Plácido*: No, no, Milanés, había que vivir. La Isla entera convidaba a vivir. Tú lo sabes. Mucho azul y mucho verde y el aire embalsamado de las madrugadas.
> *Milanés*: Vivir con decoro o enloquecer.
> [*Milanés*: I wrote that poem irritated at how you wasted your talents. (*Annoyed*.) How could you write those odes, sing the birthday of

a stupid girl, praise a fat, gouty old man with money? I couldn't understand it.
Plácido: It's very simple. I had rumblings in my belly and I needed to fill it, if not the roar would have covered the Island. (*Quietly*.) And they could accuse me of being subversive. Treason, is the exact word.
Milanés: I wasn't rich either.
Plácido: But you were white.
Milanés: One needed to be inflexible, not to cede before corruption.
Plácido: No, no, Milanés, one had to live. The entire Island invites one to live. You know that. A lot of blue and a lot of green and the balmy air of the mornings.
Milanés: Live with decorum or go crazy] (Delirio 105).

Plácido's response to Milanés reveals basic differences between the two poets: Plácido was driven to celebrate occasions that Milanés (and Plácido himself) considered to be unworthy of poetry. In effect, for Milanés, he is selling his talents to the highest bidder. Plácido, in turn, replies that he was protecting himself from hunger and from political persecution, a defense that Milanés cannot comprehend. This difference in the two reveals the primary distinction that race played (and plays) in determining lives. At different points in *La dolorosa historia*, the present moment connects to an earlier one, finding how the two can connect across the years. Here, we see one of those moments, where Estorino finds in Plácido and Milanés' conversation a link to the idea of complicity and political implication that characterized 1970s Havana. Just as Milanés cannot understand Plácido's seeming collusion, Havana's artistic community in the 1970s was characterized by questions of complicity or accusations of support of revolutionary and counterrevolutionary ideas.

For Milanés, a white man from a well-respected, though poor, family, survival meant something different than that which it meant for Plácido, an illegitimate mulatto whose only ambition could be to become a hairdresser or a carpenter. As the two men continue this conversation, Plácido tries to reveal what race meant in nineteenth-century Cuba. It is important to remember that throughout this conversation, the signs of savagery are present in Plácido's bloody body, recalling for the audience his violent end:

Plácido: Tú pertenecías al mundo, era un mundo blanco.
Milanés: En ese mundo blanco yo no pude estudiar, en ese mundo blanco fui rechazado por mis parientes, en ese mundo blanco sentí tanto asco que prefiero mi silencio.
Plácido: En ese mundo blanco tú podías elegir. Yo no. Yo era rechazado porque mi padre había sido un mulato cuarterón, y sólo podía ser: carpintero, peinetero, músico. Decidí ser poeta. Y se la cobraron. No les gustó que yo eligiera. "Qué atrevimiento el de ese mulato que no se da

su lugar y quiere igualarse a nosotros y usar el idioma castellano, blanco, como si fuera el suyo. Y además lo emplea bien y el pueblo lo aclama, lo admira, lo busca, repite lo que dice. Es demasiado atrevimiento." Y ese mundo blanco inventó una conspiración fantástica para acabar con un mundo mulato que se iba formando. Y por aquí entró la bala.

[*Plácido*: You belonged to the world, it was a white world.

Milanés: In that white world I couldn't study, in that white world I was rejected by my relatives, in that white world I felt so much disgust that I prefer my silence.

Plácido: In that white world you could choose. I couldn't. I was rejected because my father had been a quadroon, and I could only be a carpenter, hairdresser or musician. I decided to be a poet. And they made me pay. They didn't like that I had chosen. "How dare that mulatto not know his place and want to be our equal and to use the white, Castilian language as if it were his. And besides that he uses it well and the people love him, admire him, seek him out, repeat what he says. It's too much." And that white world invented a fantastic conspiracy to put an end to the mulatto world that was forming. And here is where the bullet went in] (Delirio 105–106).

This entire scene touches on many of the important issues that can be found in both the play and Cuban society, at large. First, there is an immediate emphasis on the role of race and class in both the poets' perceptions of the world that surrounds them and their reception by it. Plácido points out that his life is determined by being mulatto; his choices are limited and the people around him mark his difference and consider his literary efforts as an affront because of his race. Milanés, on the other hand, sees the similarities in limitations between the two lives and cannot understand that Plácido would abandon his talent and write poetry that was beneath him. While Milanés' experience is also limited by financial restraints, he cannot fathom how Plácido could be driven to "ceder ante la corrupción [cede before corruption]." Plácido, in turn, sees his own actions as survival, counteracting the very prejudiced beliefs that would have him executed. Here Plácido finds himself in a similarly contradictory space as that in which Jill Lane in *Blackface Cuba* identifies Juan Francisco Manzano as occupying. For Manzano, despite (or precisely because of) his achievements and his literary worth, according to a white readership, it is the very act of being a slave able to write that places him in "the position of a special exhibit, a spectacle of marvelous social contradiction."[34] Plácido is a parallel contradiction, where the act of writing sets him apart and is what saves and condemns him simultaneously, remarkable circumstances that were not binding to Milanés, given his race.

The central part of this scene and of the entire play occurs when the characters begin to speak about and use torture as a method of control—of an other, of the situation, of one's self. First, Polonia, a black slave owned by Esteban Santa Cruz de Oviedo, is dragged onto the stage where she reveals the rebellion leaders' plan to attack and kill the white men. The Governor orders an investigation and Oviedo, along with Francisco Hernández Morejón, captures a group of escaped slaves who they believe to be in on the rebellion. The entire conversation in which the white men discuss the impending uprising and their efforts to suppress it has an ironic air—their violent intentions are obvious. That is, the spectator is completely aware that the men are inventing the situation for their own professional and financial benefit with no regard for the truth or for the lives that they are about to end. Nevertheless, their words testify to the fact that they feel they must follow some protocol no matter how false these images may be.

This sardonic code of behavior continues when some slaves are captured and brought on stage to be tortured. El Español, a character who represents the colonial government in Cuba, puts the Military Commission in charge of the situation and allows it free hand to proceed at will. Three ladders are brought onto the stage and the black men are tied to them. At the same time, a priest begins to outline the following regulations on how the torturers can progress and who can and cannot be tortured:

> La tortura no puede hacerse hasta ocho horas después de haber comido y esto para que no se conturbe el estómago, vomite el reo, y le sobrevenga enfermedad grave e incurable. No se puede torturar a menores de catorce años ni a mayores de sesenta y cinco años; a los que padecen de fiebre, apoplejía, epilepsia o "Gravis morbo gallicus," a los que han sufrido graves contusiones en la cabeza, garganta, pecho, vientre, brazos; a los corpulentos por superabundancia de grasa, a los estrechos de pecho, monstruosos, gibosos, desiguales de brazos y mujeres embarazadas.
>
> [Torture cannot be conducted until eight hours after eating, so that it doesn't disturb the stomach, the offender doesn't vomit, or a serious and incurable disease doesn't come over him. Neither those younger than fourteen nor those older than sixty-five can be tortured; nor those who have fever, apoplexy, epilepsy, or " gravis morbo gallicus," nor those who have suffered serious contusions to the head, throat, chest, stomach, arms, nor those who have excess fat, are thin in the chest, have monstrosities, hunchbacks, uneven arms, or pregnant women] (Delirio 109).

The priest ironically outlines certain regulations that govern when torture can be used and against whom in order to uphold some sort of "humane" interpretation of torture and its role in the community. First, his rules leave only a small window of time in which it is possible to torture, and then many possible victims are made exempt by his decree. Thus, according to this, torture becomes highly regulated and the church is cleared of involvement in the episode. Nevertheless, the *Fiscal* responds negatively to the rules outlined above and states that the power of the state to protect itself and its citizens is stronger than any other rules that exist:

> Como presidente de la Comisión Militar encargada de esclarecer todo lo concerniente a la conspiración de los negros contra la raza blanca, declaro: cuando se trata de la seguridad del país y de un delito de Estado, cualquier medio es legal y permitido si de antemano existe la convicción moral de que ha de producir el resultado que se desea y es exigido por el bien general.
>
> [As president of the Military Commission in charge of everything concerning the conspiracy of the blacks against the white race, I declare: when we are dealing with the security of the country and a crime of the State, whatever method is legal and permissible if beforehand there exists the moral conviction that it will produce the desired and demanded result for the general good] (Delirio 110).

In this way, the church is stripped of its authority before the security of the state, and the government is given *carte blanche* to execute its own orders and desires to quell any possible uprising.

After having discussed the legitimacy and legality of torture in the proceedings, the *Fiscal* actually presides over the torture of a black slave onstage. This is a particularly important scene, not only because of the savagery that it portrays, but also for the hypocrisy to which it attests. In this section of the scene it is only the *Fiscal* and a black slave who is named Negro 1 who speak, although there are others onstage who perform the actual beating on the body of the slave. This is an interesting split between the *Fiscal*'s statements and the anonymous men's actions, suggesting that the words and the acts are being separated in an effort to divorce one from the connotations of the other. The *Fiscal*'s interrogation of Negro 1 consists of him asking for information on the reported uprising against the white slaveholders. When Negro 1 doesn't give the "correct" answer (whether this is by choice or because of his own lack of information is unclear), the *Fiscal* orders him to be whipped, waiting to hear the answer he is expecting:

> *El Fiscal*: Di la fecha, la fecha del levantamiento. (*Lo azotan.*)
> *Negro* 1: Nochebuena.
> *El Fiscal*: ¿Estás seguro?
> *Negro* 1: Pascuas, será en Pascuas.
> *El Fiscal*. Ladino, esos errores ocultan tu obstinación. ¿Cuándo, cuándo?
> *Negro* 1: Nochebuena, Nochebuena Chiquita.
> *El Fiscal*: Fijaos cómo trata de evadir la investigación de este tribunal. Ni el tormento es capaz de hacerle abrir su corazón. Si en dicho tormento

muriese o fuese lisiado, sea a su culpa y cargo y no a la nuestra, por no
haber querido decir la verdad. (*Lo azotan*.) ¿Es cierto que se pretende
acabar con la raza blanca, quemando los cañaverales, matando el ganado
y sorbiendo su sangre?
Negro 1: Sí.
[*El Fiscal*: Give the date, the date of the uprising. (*They whip him*.)
Negro 1: Christmas Eve.
El Fiscal: Are you sure?
Negro 1: Christmas, it'll be at Christmas.
El Fiscal. Ladino, these errors hide your obstinacy. When, when?
Negro 1: Christmas Eve.
El Fiscal: Look at how he tries to evade the investigation of the tribunal.
Not even torture is enough to make him open his heart. If in this torture
he were to die or to be crippled, it would be his fault and not ours, for
not wanting to tell the truth. (*They whip him*.) Is it true that they want
to finish off the white race, burning the plantations, killing the livestock
and drinking their blood?
Negro 1: Yes] (Delirio 110–111).

The *Fiscal*'s prompting during the interrogation demonstrates the pattern that will follow in the next scene of questioning and torture. The slave wavers in his answers, either demonstrating that he does not really know the answers or that he is unwilling to give them up so easily. When he does answer with what he thinks will be more acceptable, the *Fiscal* questions him and tests the answers with more physical violence, as if he were, in the words of Page duBois, a touchstone that only reveals the truth under torture.[35] This scene recalls duBois's study of torture in ancient Greek society, where a slave was only believed to be telling the truth when that information was extracted through torturing the slave's body.[36] In this way, the *Fiscal* and the government that he works for are absolved of the crime of exacting any injustice because torture is the only way to guarantee truth. Furthermore, the *Fiscal*'s last words—that if the slave should die or be injured from this torture, it will be his own fault for not wanting to tell the truth—brings the spectator-reader back to the *Fiscal*'s response to the priest's prohibition of torture and attempts to release him and his men from any potential wrongdoing. This statement is interesting, given that it acts simultaneously as a verbal, and public, absolution and as a warning to the slave (and to others) of what could happen to him if he doesn't answer "truthfully" (read: with the answers that the *Fiscal* wants).

Following this admonition, the *Fiscal* begins to ask for the names of both the black and white men who are leading what he sees as an impending revolt. The slave, who doesn't know for whom his interrogator is looking, is fed the answers by the *Fiscal* himself, repeating back names or pseudonyms of the condemned men. With his utterance, he seals the men's fate and validates the *Fiscal*'s authority over him and the others:

> *Negro* 1: No sé los nombres.
> *El Fiscal*: Jorge López.
> *Negro* 1: Jorge López.
> *El Fiscal*: ¿Es Jorge López un enemigo de la raza blanca?
> *Negro* 1: Sí.
> *El Fiscal*. ¿Es Santiago Pimienta un enemigo de la raza blanca?
> *Negro* 1: Un enemigo de la raza blanca.
> [...]
> *El Fiscal*: Ya estás en buen camino, dime el nombre de quien dirige toda esta intriga.
> *Negro* 1: Dime el nombre.
> *El Fiscal*: Tú lo sabes. Es un mulato que escribe poesías.
> *Negro* 1: Un mulato que escribe poesías.
> *En el cortejo se oyen voces que gritan*: "*Huye, Plácido.*"
> *El Fiscal*: Plácido.
> *Negro* 1: Plácido.
> [*Negro* 1: I don't know the names.
> *El Fiscal*: Jorge López.
> *Negro* 1: Jorge López.
> *El Fiscal*: Is Jorge López an enemy of the white race?
> *Negro* 1: Yes.
> *El Fiscal*: Is Santiago Pimienta an enemy of the white race?
> *Negro* 1: An enemy of the white race.
> [...]
> *El Fiscal*: Now you're doing well, tell me the name of who's directing this intrigue.
> *Negro* 1: Tell me the name.
> *El Fiscal*: You know it. It's a mulatto who writes poetry.
> *Negro* 1: A mulatto who writes poetry.
> *In the procession voices are heard that shout*: "*Flee, Plácido.*"
> *El Fiscal*: Plácido.
> *Negro* 1: Plácido] (*Delirio* 113–114).

This exchange between the *Fiscal* and Negro 1 uncovers the reality behind the justice that was measured out. The interrogation consists of torture and a set-up that implicates the people that the *Fiscal* already suspects as the culprits, thus robbing the whole situation of any value as justice. Estorino, in this way, reveals that the justice of this historical moment was anything but, and forces the spectator to take another look at the stories handed down in the history books and question their validity. Simultaneously, Estorino inspires a questioning of other moments that have been tainted by "interrogation," such as that of Heberto Padilla in the 1970s.

When Plácido is dragged onto the stage to face the charges against him, the *Fiscal* verbally lists these 'crimes.' These accusations, as quoted below, include actions that would not be thought to be against the law but attest to Plácido's "suspicious" behavior in a twisted situation that views the poet as intending to finish off the white race and to, conversely, proliferate the black one. Nevertheless, regardless of how ridiculous the charges may sound, they serve to condemn Plácido to death:

> Las pruebas contra usted son contundentes: viaja por la Isla haciendo contactos con los juramentados; ha escrito un poema donde habla de crimen y sangre; es un individuo muy peligroso, vago e inútil. Y como prueba concluyente se ha casado con una negra. Queda comprobado que sus propósitos son acabar con la raza blanca y por lo tanto será fusilado por la espalda.
>
> [The proof against you is forceful: you travel the Island making contacts with the oath-takers; you have written a poem where you talk about murder and blood; you are a dangerous, lazy, useless individual. And as conclusive proof you have married a black woman. It is proven that your purpose is to finish off the white race and for that reason you will be shot in the back] (Delirio 115).

Just like the interrogation of Negro 1 quoted above, the *Fiscal*'s evidence against Plácido is all just a façade that lets it be known that the justice that the colonial government meted out was not justice at all, but a twisted set-up that reproduced unjust conditions. Estorino aims to un-do the historical spinning that he sees as creating the Conspiracy of the Ladder as it has been historically told. His purpose is to re-claim the poet and his work (here Plácido) and a debate on the treatment of dissent and remind the twentieth-century spectator of the extremes to which an unjust authority can go.

On this topic, one can see a parallel with Estorino's own era in that Heberto Padilla had undergone a similar interrogation and accusation just three years before *La dolorosa historia* was written. In this way, the blurring of time periods between Milanés' and Estorino's continues, as Estorino highlights a connection across the century. His allusion to the ills of slavery and of the censorship of dissenting voices during the abolition movement portrayed in *La dolorosa historia* simultaneously refers to the censorship that he and other playwrights experienced in the early 1970s, making it virtually impossible for innovative theater to be produced.

After Plácido is condemned, Milanés occupies center stage once again with the Mendigo. The two men discuss the future actions that Milanés has decided to undergo: he has chosen to be tortured in order to experience what that means. In a scene that recalls Milanés' reaction to the earlier bloodshed, the nineteenth-century poet places himself on the ladder to be interrogated and tortured by Negro 2 in order to understand what torture is:

> *Negro 2*: ¿Por qué escribiste contra la esclavitud?
> *Milanés*: Porque tenía ideas humanistas y no podía soportar la crueldad de unos hombres contra otros.
> *Negro 2*: Si no has probado el látigo no sabes lo que es crueldad.
> *Milanés*: Conozco otra crueldad. Yo había sido humillado.
> *Negro 2*: Nosotros también, pero hasta un extremo que tú no eres capaz de imaginar.
> *Milanés*: Aquí estoy. Despiértame la imaginación.
> *El Negro 2 lo azota*.
> [*Negro 2*: Why did you write against slavery?
> *Milanés*: Because I had humanist ideas and I couldn't stand the cruelty of man against man.
> *Negro 2*: If you haven't felt the whip you don't know what cruelty is.
> *Milanés*: I know another cruelty. I have been humiliated.
> *Negro 2*: So have we, but to such an extreme that you can't imagine.
> *Milanés*: Here I am. Awaken my imagination.
> *El Negro 2 whips him*] (Delirio 118).

Milanés and the Negro 2 recreate the torture scene—except here the tortured body is present of its own free will and the roles have been reversed, the white man is being tortured and the black man is torturing. This is an important distinction, but it is also central to the scene that the tortured poet is *choosing* to put himself on the ladder in order to understand the humiliation of public torture. This scene remembers Foucault's writings on public torture and executions, yet here it is a desire by Milanés to comprehend this suffering in order to put an end to torture.

Behind Milanés' desire to understand the humiliation of public torture lies another request. The poet wants to comprehend the bloodshed that has occurred on both sides of the rebellion and the need for more revenge and blood that lies behind the violent horrors that have been committed:

> *Negro 2*: ¿Comprendes por qué encendimos la tea?
> *Milanés*: Yo estaba contra toda la violencia.
> *Negro 2*: ¿Y cómo se lucha contra esta violencia? (*Le muestra el látigo*.)
> *Milanés*: Ahora no podrás decir que no comprendo tus dolores.
> *Negro 2*: Y entenderás nuestra violencia.
> *Milanés*: No, no, eso no. Hay que encontrar otro camino.
> *Negro 2*: El único: ojo por ojo, y diente por diente. Fueron miles los perseguidos y azotados. Los más débiles se suicidaron. Otros murieron en la escalera sin decir una palabra. Me tocó esa suerte. El rencor comiéndome por dentro, mordiéndome los labios para no gritar.

[*Negro 2*: Do you understand why we lit the match?
Milanés: I was against all violence.
Negro 2: And how do you fight against this violence? (*He shows him the whip.*)
Milanés: Now you won't be able to say that I don't understand your pains.
Negro 2: And you'll understand our violence.
Milanés: No, no, not that. There must be another way.
Negro 2: The only way: an eye for an eye, and a tooth for a tooth. Thousands were pursued and whipped. The weakest committed suicide. Others died on the ladder without saying a word. That was my luck. The resentment eating me from the inside, biting my lip so as not to scream] (Delirio 121).

Milanés questions the need for violence in order to undo past atrocities. He chooses to seek another way that does not include violent revenge as the only means to a new society. Echoed in Milanés' exchange with his torturer, the spectator can hear a similar argument about the Cuban Revolution and the official response to dissidence. Estorino's play uses the atrocity of the past century to point to the social and political atrocities that are happening once again in Cuba. Estorino, through the words of Milanés, questions whether the only possible response to the past is a violent break, such as that which we see in this scene and in the official response to dissidence such as Padilla's. Instead, Milanés and *La dolorosa historia* search for another possibility that may end the cycle of violence.

This connection to the modern day in a historical play is one that is explored extensively by Herbert Lindenberger in his *Historical Drama: The Relation of Literature and Reality*. This is a particularly interesting study for Estorino's *La dolorosa historia* for many reasons, but perhaps one of the most important at this point is the examination of what Lindenberger calls martyr plays. These are linked with tyrant plays in his description but the former detail plays where the emphasis is placed on the victim rather than the perpetrator of crimes. Martyr plays have a certain connection with an imitation of Christ and focus on the martyr's inner development, an aspect that can make them less theatrical.[37] These observations are important to consider in connection with *La dolorosa historia* in that it can be considered a martyr play, portraying Milanés in the role of a martyr of the tyranny of the nineteenth century and is, thus, an admonition against what could happen again. Though rather than make martyrdom appealing, I believe that by connecting his own moment with that of the past, Estorino aims to point to the sins of the nineteenth century in order to avoid their same occurrence in the 1970s.

La dolorosa historia del amor secreto de José Jacinto Milanés is the dramatic re-telling of the life of the nineteenth-century poet and playwright José

Jacinto Milanés and the historical context in which he lived. Nevertheless, Abelardo Estorino's play does much more than just re-tell Milanés' life; Estorino creates a new Milanés that will allow him to focus attention on the man's life and his literary work while simultaneously referring to the context of Cuba in the 1970s. By compelling the reader-spectator to re-discover this important Romantic poet, Estorino questions the accepted beliefs that have been passed on about Milanés and his literature and he forces the reader-spectator to question *their* accepted beliefs on multiple conventions through his own dialogue on Milanés and his work. By integrating the nineteenth-century poet's work into his own dramatic production, the borders between genres and autonomous works blend and begin to disappear as do the borders between the centuries. With this, the abolitionary and independence movements of the nineteenth-century are seen to offer insight into *La dolorosa historia*'s revolutionary context of social upheaval. Estorino highlights onstage how the construction of history and memory is etched into our thoughts and stories through the use of violence, betrayal and racism. In this way, *La dolorosa historia del amor secreto de don José Jacinto Milanés* connects across centuries, literary genres and memories in order to encourage a re-writing of the past. Historical re-interpretation is a violent gesture that revisits questions of class, race, gender, and literature. The violence of sexuality, class, and gender will take center stage in Eduardo Pavlovsky's exploration of his own time period, as we will see in the next chapter. Despite the differences in time period, Estorino's and Pavlovksy's plays share a similar view. Whereas *La dolorosa historia* returns to the nineteenth century to explore issues of silencing and freedom, Pavlovsky's *La mueca* is a contemporary snapshot of the social and political context in which it is written. This play examines bourgeois hypocrisies alongside a look at the increasing breakdown in social order that characterized the 1970s in Argentina. Just as in Piñera's *Dos viejos pánicos*, Pavlovsky chooses to explore his own historical moment in an almost absurd look at what is happening. The fundamental theme in this play, like the two Cuban plays that we have analyzed up to this point, is the role of violence in the making of human relationships. Just as how violence bound Tota and Tabo to one another and how violence defined the Cuban nineteenth century, violence in the 1970s in Argentina will characterize how people look at one another and how they understand themselves in this world.

Chapter 3

Filming the Bourgeoisie: Defining Identity with Violence in Eduardo Pavlovsky's *La mueca* (1970)

Hoy no hay guión.

Eduardo Pavlovsky, La mueca

La mueca (1970) by the Argentine Eduardo Pavlovsky (b. 1933) portrays the tale of four intruders-filmmakers who decide to confront a bourgeois couple about their hypocrisies, and in the process bring to light the role of performance in the couple's construction of identity. Whereas in Piñera's *Dos viejos pánicos* violence was used to *cover up* a deep-rooted fear, Pavlovsky's violence *inspires* fear in its victims in order to humiliate them and make them realize the precariousness of their situation. Violent gestures are used to define the characters and allow them, in turn, to define others. Violent laughter that interrupts actions or mocks another, for example, becomes a tool that permits the one who laughs a place of authority—an authority that is born from humiliation and made to break someone else. Violence, then, is used in *La mueca* to create and destroy relationships and to form identities and self-definitions through the use of force.

The self-construction of identity as well as the constructions of others' identities is a central task of the characters in *La mueca*. Class is one of the tools through which the actors try to define themselves and others. Social class here is an instrument that plays a role in defining both one's self and others as well as the subsequent relationship that will emerge between people—in other words, it creates identity. However, these identities that are made are constantly called into question through the characters' words and actions. Identity, then, becomes a fluid quality that the characters can transform depending on the circumstances in which they find themselves. Similarly, naming in *La mueca* mirrors the confusion found in identity in that Pavlovsky gives his characters names that attempt to define them socially to a certain role. However, like identity, they seem to slip away and Pavlovsky purposefully highlights their contradictions and failures to adhere to an identity. They are but distortions, and their names underline the lack of connection. Like the very title—*La mueca*, the grimace—the characters are distortions of what they 'should' be. Pavlovsky proposes that the role that each character takes on is a performance in the theatrical sense that tries to connect to an original but can never quite achieve it. That is, each character tries to play a part rather than portray his/her own personality. The hypocrisies, then, that are revealed throughout the play are inevitable in a world of pretending and simulation.

The role of performance that can be seen in the characters, and as a central theme in the play, highlights the role of metatheater in *La mueca*. Pavlovsky invites the play to turn against itself and question its surroundings. The seams of what makes theater are exposed by the playwright in an attempt to look at, and see how, the violent constructions that we push ourselves into ripple through our own actions. The fact that the play, in turn, concerns the making of a film underlines the desire to see more clearly the violent acts that construct our identities. The correlation between film and theater that are put forth highlights the role of voyeurism within the theatrical representation. As Matías Montes-Huidobro acknowledges, the play is about a voyeurist making a film with other voyeurists and finally turning the other characters and even the spectators into voyeurists.[1] These multiple layers show the level of performance that allows the examination of the intersections between the construction of identity and violent acts in theater. *La mueca* is not simply chronicling an intrusion, but endeavors to push the spectators to understand and question the role of performance in our own lives. The play both connects and transcends its geographical context, as will be seen.

Buenos Aires has long been considered a theatrical center in Latin America and Eduardo Pavlovsky is one of Argentina's principal dramatists of the second half of the twentieth century. Born in 1933 in Buenos Aires, Pavlovsky studied to become a doctor, specializing in psychoanalysis, while he simultaneously pursued an interest in theater. His attraction to theater was originally limited to acting (and, in fact, he continues to act in most of his plays), though he quickly became involved with writing plays, starting with his 1962 *Somos*. In the 1960s, like many Latin American playwrights of the time period and especially Piñera and Gambaro, he was influenced by the European and North American playwrights and theorists associated with the Theater of the Absurd, although Pavlovsky has preferred to refer to it as a theater "hacia una realidad total [toward a total reality]."[2] For Pavlovsky, the vanguard theater that was being written and produced at the time—usually referred to as the Theater of the Absurd—was a theater that searched for meaning:

> Personalmente entiendo el teatro de vanguardia como un teatro primordialmente de búsqueda. Es un teatro que dice "hacia" y que intenta conectarse con aspectos absolutamente "reales" de nuestra personalidad. Es el teatro de nuestros cotidianos estados de ánimo. Aquél que nos libera de los grandes discursos y mensajes, es el teatro de nuestras eternas preguntas incontestables, el teatro de nuestra soledad, el que nos hace hablar en el idioma de nosotros, *con nosotros*, y no de nosotros, *con los otros*
>
> [Personally I understand avant-garde theater as fundamentally a theater of searching. It is a theater that says "toward" and that tries to connect with absolutely "real" aspects of our personality. It is a theater of our quotidian moods. That which liberates us from the grand discourses and messages, it is the theater of our eternally unanswerable questions, the theater of our solitude, the one that makes us speak in the language of us, *with us*, and not about us, *with others*.][3]

Theater, for Pavlovsky, has to reflect the present, and in this way, it is the theater of contemporary man, taking in all the doubts and uncertainties embodied in him.[4] The purpose behind his involvement in the theater, then, is to move the audience to think and to reflect upon the world in which they move. It is a conversation, in his opinion, not a monologue. The role of collaboration seen here is of particular importance to Pavlovsky. He, more than the other playwrights studied here, emphasizes the fundamental role of collaboration in theater.

The 1970s brought about a change in Argentine and, most importantly, in Pavlovsky's theater. Jorge Dubatti, a noted critic, identified changes in Argentine theater during the period of 1968 to 1976 and the beginning of the dictatorship and military government, referred to in Argentina as the Proceso de Reorganización Nacional (1976–1983; in the United States it is often called the Dirty War). At this time, in what Dubatti categorizes as a "segundo momento [second moment]" of the decade of the 1960s, Argentine theater became more political.[5] This shift corresponds with many other artistic currents in South America and other parts of the world after 1968. Pavlovsky's theater, in particular, beginning with *La mueca* in 1970, stressed a preoccupation with examining and identifying "el momento histórico-social [the historical-social moment],"[6] an intellectual concern which culminates in *El señor Galíndez* (1973), perhaps the play for which Pavlovsky is most well-known. Pavlovsky's interest in political topics continued throughout the 1970s and he was forced to leave Argentina in March of 1978 when paramilitary forces tried to kidnap him while he was presiding over group therapy.[7] According to his own accounts, he escaped through a window and exiled himself to Spain until 1981.[8] He returned to Buenos Aires and formed an important part of Teatro Abierto, a theater movement against the military dictatorship, and he remains to this day an essential voice in Argentine and Latin American theater.[9]

La mueca was first published in 1970 by Casa de las Américas in Havana in the collection *Tres obras de teatro*, having won an honorable mention for theater in the same year. It was accompanied in this collection by *El avión negro* by Roberto Cossa, Germán Rozenmacher, and Carlos Somigliana and *Flores de papel* by Egon Wolff. The play premiered in Buenos Aires with the Grupo de Actores Profesionales (G.A.P.) in the Olimpia on 12 May, 1971, directed by Oscar Ferrigno, with Pavlovsky himself in the role of Carlos. This level of self-commitment sets Pavlovsky's theater apart from that of many other playwrights in that he takes on an integral role in the production of the premiere of his plays—often the role of one of the characters. While a shared and creative process is fundamental to the theatrical genre, Pavlovsky emphasizes in his writings and interviews more strongly than others the role of the collective process in the ultimate product that comes from his original script. He believes in the centrality of the collaboration of various readings by the author, the director, the actors, and the audience, in creating the final product that is the produced play on stage.[10] In this way, considering Pavlovksy's own investment in the production of each of his plays and his extensive role in the continued process, his theater must be considered differently than that of the other playwrights studied here. Whereas for the other three plays, their possible production must be imagined with very little confirmation, *La mueca* has specifically undergone this process and has, thus, reached the culmination that was first imagined by its original author.

Eduardo Pavlovsky's *La mueca*, written in 1970, marks a significant period in Argentine history. The country continued to be defined in many ways by the government of Juan Domingo Perón, who had been removed from office in a coup d'etat in 1955. The period following this is seen to be successive attempts to try to fix the economic and political organization that was overthrown by a coalition of military leaders and liberals. The country suffered from chronic inflation and social and political difficulties. Despite Perón not physically being in Argentina, his presence was felt through the continued influence of Peronism on politics. General Aramburu, president from 1955 to 1958, took a hard line against Peronism which incited resistance. These pressures contributed to Arturo Frondizi's election as president in 1958 and also to his eventual ousting in 1962. Initially, Frondizi appealed to the Peronists by being pro-union, despite the ban on the former in the elections. He secretly won the exiled Perón's endorsement and, consequently, the elections. This connection with Peronism provoked the Army's suspicions and Frondizi was forced to walk a fine line between the two throughout his administration. However, by March of 1962, this had collapsed. Perón had publicly broken with Frondizi in June 1959 and Frondizi's connections to Cuba alienated the Army. In the congressional and gubernatorial elections of 1962, the Peronists carried the majority of provinces and the Army urged Frondizi to annul the elections. When he refused, he was deposed and arrested. José María Guido was installed as a puppet president until elections were held in 1963. Arturo Illia was then elected president and the government attempted another solution to the political instability in the wake of 1955.

Illia, however, was deposed in 1966 when it became clear that he did not have the solution to the stagnant economy. General Onganía, who had recently resigned as Chief of Staff of the Army, became president in a coup that ushered in a decidedly authoritarian administration that consolidated power in the hands of the president. He rejected the title "provisional" that other coups had used, marking a difference between this one and others that had come since 1955. Initially, there was some progress and stability both economically and politically as the new minister of economy, Adalbert Krieger Vasena, implemented his plan and as Onganía took advantage of divisions among the unions and the Peronists. However, his limiting of university autonomy and moves against unions exploded in 1969, most specifically in what has come to be known as the *cordobazo*. In May 1969, demonstrations of students and automobile workers erupted in the city of Córdoba and quickly turned into rioting as others joined them. These 48 hours of riots ended the temporary dormancy that Argentina had been enjoying and divided the army. Onganía wanted to respond with a show of force, whereas General Alejandro Lanusse wanted concessions. The latter opinion won out and Onganía replaced his cabinet. However, the ensuing months were marked by violence, which would only increase as the country moved into the 1970s.

As the government's control weakened, armed struggle became more and more prominent. Leftist groups, such as the Montoneros, were matched by others on the right. These groups participated in various violent acts, the most famous of which was the kidnapping and assassination of the ex-president Aramburu in March of 1970. This action proved to be the end of Onganía's administration. General Roberto Marcelo Levingston became president with Lanusse in the background. Other political groups banded together and called for a resumption of civilian rule in a manifesto they called the "Hora del Pueblo [Hour of the People]." In February 1971, more rioting broke out in Córdoba and the following month Levingston was replaced with Lanusse. Lanusse attempted to win the support of the unions, but inflation rose. In July 1972, with the *Gran Acuerdo Nacional*, he called for all democratic forces to unite against subversion, in an effort to gain control. Finally, Lanusse lifted the eighteen-year ban on Peronism.

Perón had been in exile in various different countries in Latin America and was in Madrid at this time. Nevertheless, he had maintained close ties with groups in Argentina through letters and emissaries. He had been manipulating the various Peronist groups against one another, so that, at this time, Peronism was, in the words of historian David Rock, "all things to all men."[11] Perón returned to Argentina in November 1972 for the first time since 1955. With his support, Héctor Cámpora won the elections but would only remain in office for 49 days. Perón was again to return to Buenos Aires in June 1973. However, the crowds waiting for him at Ezeiza airport broke into fighting and provoked numerous deaths. This episode highlighted Cámpora's inability to govern and Perón withdrew his support. In the new elections held after the unseating of Cámpora, Perón won and he returned triumphantly to Argentina on October 17, 1973.

Perón's return was a sign of how desperate the military was to contain the left.[12] Initially, it did seem that there was some hope and excitement for the future, though there was little decrease in the violence that had so marked the previous years. In late 1973 the Montoneros and the ERP (*Ejército Revolucionario del Pueblo* [Revolutionary Army of the People]) continued their kidnappings, assassinations, and robberies in an effort to win money and control of the political situation. Now, however, their activities were equaled on the right by a new secret organization, *Alianza argentina anticomunista* [Argentine Anticommunist Alliance], known as Triple A and said to be connected to the Federal Police.

Perón had chosen his third wife, María Estela Martínez de Perón, "Isabel," as his running mate in an effort to manage the differing groups with which he had been in talks by not choosing any of them. Upon his return, economically, Argentina enjoyed a calmer moment, which he tried to encourage politically. He did this by trying to eliminate other voices and strengthen his own control over the party. This would entail getting rid of the Peronist groups that helped to return him from exile, such as the laboristas, the Montoneros, and the Juventud Peronista. The attack on terrorist groups (which conveniently ignored the activities of the Triple A) highlighted how he had used the left and how they had allowed it. In May of 1974 he broke publicly with them, an act that was followed by his death of a heart attack on the first of July of that year.

Isabel Perón ascended to power following Perón's death, but she was not equipped to deal with the violence and economic difficulty that marked 1974 and 1975. Guerrilla warfare continued and the Army intervened in 1975, after the assassination of the Chief of Police. There was talk that Isabel Perón had stolen money and impeachment proceedings were held against her. This rising violence, apparent corruption, and the surging inflation all contributed to a situation that was ripe for a coup d'etat and on March 24, 1976, the Army abducted the president and the Proceso de Reorganización Nacional, what is known as the Dirty War, began.

La mueca was written in 1970 and premiered in 1971, soon after the *cordobazo* but before Perón's return. This was a time marked by violence, both clandestine coming from left- and right-wing groups and political as the government attempted to control the various responses. It was not just the immediate geographical context that helped to define *La mueca*, but also the more global surroundings of the worldwide student demonstrations, the Vietnam War, and the drug counterculture. Pavlovsky's play contributes to and questions this context in an effort to awaken the spectators and the participants to the violence that defines relationships and threatens to undo lives. In this way, the play is both tied to and transcends the historical moment and geographical context to which it contributes.

Pavlovsky's *La mueca* is divided into two acts and portrays the encounter between four men of the underworld and a bourgeois couple. The purpose of the men's intrusion into the apartment reveals itself slowly over the course of the first act, divulging bit by bit that their plan is not to rob the house as spectators may assume, but to film a movie. In the first act, the audience sees the four characters (Sueco [Swede], Aníbal, Turco [Turk], and Flaco [Skinny]) enter Carlos and Helena's empty, bourgeois house. In the beginning, Sueco seems to be in charge and orders Aníbal and Turco to lock the door and thus, shut out Flaco, who has stayed behind buying cigarettes. As we discover over the course of the play, Flaco comes from a different social class and threatens Sueco's control of the group. In fact, control over the group seems to be a constant source of friction in that the men, especially Sueco and Flaco, need to manifest at least for a moment their ability to dominate the others.

The four men empty their bags in the second floor of the house while the purpose of the men's visit (to make a film) is revealed bit by bit. When Flaco asks Sueco if he is reading the script, their dialogue makes known that they are there to put together a spectacle of some sort, which will today be improvised.[13] At that moment, the couple, identified at first as "Él [He]" and "Ella [She]," enters and the intruders, who have hidden within the apartment, begin filming their actions. The couple argues and, then, begins what seems to be their "mating ceremony," consisting of an odd practice where they replicate the sexual rituals of animals. Pavlovsky here is not interested in showing sex onstage—the couple goes behind a chair in order to participate in an "absurd" sexual act. At this point, the intruders come out of their hiding spaces in order to better tape the actions. Their dialogue shows they are satisfied with the "material" they are filming, thus ostensibly emphasizing the role of the film and not their own voyeurism. The couple (who is now identified by their names—Carlos and Helena) becomes aware of the others' presence, and Carlos begins to question how and why they are there. The trespassers act as if the house is theirs and offer the couple drinks which have been drugged. This provokes in Carlos an hysterical laugh and puts Helena to sleep, thus ending the first act.

This first act introduces the central tensions that will be explored in the second, and that characterize the historical and political context to which the play contributes. The second act opens a little later within the same night with Flaco reading a book on painting, Aníbal sleeping and Sueco pacing back and forth, while the other three are offstage. As is revealed in the men's dialogue, Turco has gone upstairs with Helena, and Carlos is sleeping in another room. After Flaco comments on the "injustice" of Turco being with Helena, Sueco sends the two men up to bring Turco down. Both Helena and Carlos enter the scene again and there is a confrontation between the interlopers and the couple over class differences. Carlos, in an attempt to demonstrate his own masculinity before that of the intruders, begins to physically attack Sueco while Helena tries to provoke her husband and to verbally incite the others. No one makes a move to defend Sueco; they simply state that these actions will only make it worse for Carlos. After Carlos tires himself out, the trespassers/filmmakers gain the upper hand again and reveal that they now want a confession from Carlos, to which he states he has nothing to confess. The men answer enigmatically: "Siempre hay algo que confesar. Siempre hay algo que se tiene bien guardadito [There's always something to confess. There's always something that's well-hidden]."[14] After the three men torture Carlos and threaten him with more violence, Carlos confesses that he masturbated the day before at the office in order to relax.

At this moment, the real purpose of the men's intrusion in Carlos and Helena's house begins to reveal itself, when Sueco highlights the roles that the couple plays: "es que ustedes [Carlos y Helena] viven intensamente sus papeles [it's just that you [Carlos and Helena] live intensely your roles]" (Act 2: 39). Sueco lets it be known that the men know all there is to know about Carlos, including his keeping a mistress—matched by the revelation that Helena also has a lover. When the couple sees no way to carry on with their earlier lives built from illusions and lies, the intruders decide that they have gained all the material that they can and leave the apartment. The play ends as Carlos and Helena stare at one another and listen to a recording of their earlier conversation with the intruders about their hypocrisy.

Within these two acts, *La mueca* shows its central topics to be how six characters construct identities and the violence inherent in keeping these identities regulated. The six characters reveal a tension within each individual and amongst them all as a group that they try to hide from one another through their use of violent acts. The hypocrisy that characterizes the bourgeois couple's lives becomes an important part of the play in that it is the main objective of the intruders' film, itself another aspect that hints at hidden layers that uncover other identities and violences. That is, the making of the film allows an examination of the duplicity found in the very definitions of the couple's lives, while the act of filming similarly turns the examining eye on who the *intruders* are. No character saves itself from questioning and what is revealed is how these characterizations are formed through a cycle of oppressing another—one person defines his/her identity by violating and attacking someone else's, creating a cycle of violence upon which identity is dependent.

La mueca, the title of the play, alludes to these topics of identity, control, class and oppression, and places on center stage the ideas of distortion and disguise. A *mueca* is defined as a "[c]ontorsión del rostro, generalmente burlesca [contorsion of the face, generally in a burlesque manner],"[15] a definition that provokes more questions when one first encounters the play, given that the word *mueca* is only used in the title of the play and not in the text itself. A *mueca* is a grimace that distorts the face. This suggestion indicates that the play will portray a space where all is not what it seems, and that distortion itself is what is valued. Charles B. Driskell, in "Power, Myths and Aggression in Eduardo Pavlovsky's Theater," shows the correlation between the film that the intruders are making and the *mueca* to which the title refers. For him, this film is a mask that can reveal or conceal, and the film is what allows Pavlovsky to explore the construction of identities. Driskell sees the film's purpose as that of taking away the mask that is covering the *couple*'s hypocrisy and revealing what lies beneath, the primary focus being on the couple but not on questioning human relationships.[16]

Nevertheless, I believe that the play has a wider resonance in terms of its intentions than just focusing on the couple and, by extension, the bourgeoisie. Instead, Pavlovsky incorporates an examination of the four intruders and the world in which they move. The *mueca* refers to the distortion of reality that *all* of the characters participate in and the need to dis-cover what lies beneath it. This becomes a significant issue in the analysis of the play when the characters examine the roles that they and the others play within their lives. The play becomes a means to analyze the roles that one plays both in bourgeois society but also outside of it (the four intruders are also playing a part, though they acknowledge this). This role-playing or performance is natural in theater given its transformative qualities, as Richard Schechner points out in his *Performance Theory*. Here Schechner speaks about the importance of theater "as a way to experiment with, act out, and ratify change."[17] This is significant for *La mueca* in that the characters are playing and perfecting roles to such an extent that the *mueca*, or role, they play begins to define their existence. Nevertheless, implicit in the idea of the *mueca* is that this distortion is not all positive and that the characters' transformation has a corruptive quality. This corruption lies in the difference between the couple and the intruders and their own recognition of performance in their existence. Whereas the intruders are aware of the roles that they are playing at the present moment, Carlos and Helena seem to be oblivious to this, as Sueco points out: "son fundamentalmente actores sin saberlo [you are fundamentally actors without knowing it]" (Act 2: 45). Whereas in this scene, Sueco highlights the couple's performance, Pavlovsky is underlining both that of the couple and of the intruders and, by extention, the spectators. Pavlovsky's play attempts to highlight the performance that defines the characters and the violent acts that each one uses in this definition. In this way, *La mueca*'s emphasis on performance is much more than an examination of the experimentation of theater. It is a way to advocate for change in human definitions and relationships.

Inherent in this distortion of reality is the idea of a grotesque variation on the face that covers what lies behind it, thus hiding the reality behind the performance. Delving further into the idea of a *mueca* within this play, it is important to see that the distortion that defines a *mueca* is exactly the object of this script, in that Pavlovsky examines how each character's performance is distorting a distant ideal of who they should be. The *mueca*, then, becomes a symbol of the role that each one takes on within the play, the film and within their lives. The use of this symbol in the play's title underscores the role that meta-theater will have in Pavlovsky's argument while also highlighting the omnipresence of performance in the lives of the play's characters, and, thus, in those of the spectators. The play underlines how meta-theater will allow for a certain distancing that will provoke questioning and probing of difficult topics. Furthermore, the definition quoted before from the RAE implies that the face's expression is distorted with the goal of creating a grotesque air. The scene is set for a story that promises to distort the image as well as ridicule it and the spectator knows from the beginning that the play is meant to question and provoke, revealing only the omnipresence of performance in these characters' lives.

This distortion can further be seen in Pavlovksy's use of film in that it can be seen as a distortion of theater. We must remember the connections between the two genres: the role of performance, the use of a script, the emphasis on scenery and costumes, the importance of collaboration between various artists. However, there are differences that allow us to consider film to be a distorted theater where things are similar but not completely. Just as a grimace is a distortion of a face, film guides the spectators' eyes to one place rather than emphasizing the freedom of vision found in theater. Similarly, the repetition inherent in seeing the same performance with every viewing of a film contrasts with the unpredictability that is characteristic of a live theatrical event. In this way, the movie that the intruder-filmmakers are making can be seen as a distortion of the very theatrical play in which they are participating, echoing the title and deepening its impact for *La mueca*.

Probing further the importance of names, in *La mueca*, Pavlovsky chooses to use nicknames for half of his characters: Sueco, Flaco, Turco. Even the bourgeois couple is first referred to as Él and Ella, only becoming Carlos and Helena after they are interrupted by the four intruders. What is hidden behind these names is the generic group quality that they all seem to represent rather than an in-depth individuality. George O. Schanzer, in his analysis of Pavlovsky's avant-garde theater, quickly places them in the following groupings: "el cruel Sueco, que dirige la película, el más fino Flaco, el tonto Turco, y el subalterno Aníbal [the cruel Sueco, who directs the movie, the finer Flaco, the stupid Turco, and the subaltern Aníbal],"[18] whereas the couple is decidedly categorized as bourgeois. Schanzer states that these characters' names and nicknames are not symbolic at all, a statement with which I do not entirely agree.[19] While the reasoning behind the characters' names can hint at physical attributes, the play is suggesting something more by choosing these specific nicknames and names. For example, as stated before, though the bourgeois couple has proper names, neither the man nor the woman is referred to with these names until they engage in conversation with the four intruders, stating that their actions should be seen as universally those of a bourgeois couple without the specificity attached to the individuals, Carlos and Helena. The use of the name Helena with the silent H also sets her up as different and more aristocratic than the more conventional Elena would. Similarly, the intruders have generic nicknames that hide their individual identity. Schanzer points out that a success of Pavlovsky's theater in the 1970s was his ability to capture the local (the Argentine), particularly through the characters' language, while referring to the universal through the generic quality of his characters and the action.[20] This is, in my opinion, where the importance of the characters' names lies. The nicknames indicate a loss of individuality—they become one of many—and suggest a universality that goes beyond the Argentine context evoked by their language and indicates a generality to which the action and the characters are referring. Pavlovsky is implying both his national circumstances as well as the more general environment that he sees outside of Argentina and, thus, his theater manages to point to both the local and the general.

Like the use of names, the larger question of filmmaking and its meanings is at the center of the play and needs to be considered. In the first act, the four men begin filming when the couple enters the apartment, culminating with the couple's sexual exchange. The film begins at the same time the bourgeois couple are given proper names. The four men are voyeurs of the couple's sexual encounter and they are using the camera to capture the action. The intruders' actions invite, in turn, the audience to join in and become voyeurs of the couple's actions, whether they wish to or not. Matías Montes-Huidobro, in "Psicoanálisis fílmico-dramático de *La mueca*," points out the layers and depth of voyeurism that exist in this particular scene:

> *La mueca* es una obra sobre un voyeurista (el Sueco), filmando una película en una obra donde él es uno de los voyeuristas. A su vez, los actores que se dejan ver, terminan como voyeuristas de sus propios actos y oyen la grabación de un texto que ha hecho que nosotros, los espectadores, seamos los voyeuristas que el voyeurista quiere estimular, desconectar y sacudir
>
> [*La mueca* is a play about a voyeurist (the Sueco), filming a movie in a play where he is one of the voyeurists. In turn, the actors that are being seen, end up as voyeurists of their own actions and hear the recording of a text that they have made that we, the spectators, are the voyeurists that the voyeurist wants to stimulate, disconnect, and shake up].[21]

According to Laura Mulvey's analysis of psychoanalysis and film in her seminal 1975 essay "Visual Pleasure and Narrative Cinema," there are two primary pleasurable structures of cinema: the pleasure of looking at an object as well as the narcissistic aspect of identifying with the image.[22] In the former, there is a separation that allows the audience (here, both the four intruders as well as the actual theatrical audience) to become voyeurs of the action—watching, and participating, from the outside.[23] This difference is accented, as Mulvey indicates, by the contrast of darkness surrounding the audience and the light that engulfs the image.[24] In *La mueca*, the four intruders become voyeurs of the couple's sexual encounter by intruding into their space to watch from the outside—they do not form part of the actual act but participate from the "outside" without the couple's knowledge. Through the actions of the men, the spectators, in turn, become voyeurs of the couple's sexual act—they are put in a similar position to that of the men by following their line of vision—but more removed and, thus, more of a voyeur. Similarly, the change in light between the footlights and the onstage lights that highlight the actors and the offstage darkness in which the audience is traditionally left heightens this role for the audience.

The latter form of cinema's pleasurable structure identified by Mulvey, that of identifying with the cinematic image, should also be considered when examining this scene. Here, what is most important is that the spectator (again, the four men as well as the play's audience) look at the image seen before them and recognize themselves:

> his recognition of himself is joyous in that he imagines his mirror image to be more complete, more perfect than he experiences his own body. Recognition is thus overlaid with misrecognition: the image recognized is conceived as the reflected body of the self, but its misrecognition as superior projects this body outside itself as an ideal ego, the alienated subject which, re-introjected as an ego ideal, prepares the way for identification with others in the future.[25]

According to Mulvey, the spectator (again, the intruders and the actual audience) sees himself (I can say *him*self here since Mulvey here analyzes the violence of the male gaze in cinema and in *La mueca* the intruders are male) mirrored in the image before him. In the above scene, the intruders abandon their hiding places, hoping for a better view of the couple and clearer footage for their film, understanding that this can be one of the more successful images. Mulvey's interpretations of film are helpful at this point in the play because they offer a way to comprehend the construction of identity and seeing that filmmaking has and lends to *La mueca*. Mulvey recognizes an inherent violence in film that can be seen in the way that the intruders violently trespass on the bourgeois couples' privacy and intimacy. They break into the apartment and the couple's lives in order to impose their own desires.

When analyzing this play, we must remember that the whole purpose behind the men's actions is to create a film—a sort of documentary that, though improvised, chronicles the life of a bourgeois couple in order to unveil their hypocrisies and idiosyncrasies. At the same time, the behind-the-scenes secrets of both the couple and the intruders (or, the filmmakers) are revealed. In this way, the play shows itself to be a staging, organized and managed by the director, Sueco. As with any film, the spectators see what the film wants them to see; that is, their experience is determined by the filmmakers given that they can only observe that which forms part of the film. With this in mind, it is helpful to return to Laura Mulvey's essay on film and the male gaze in order to understand this construction and Pavlovsky's use of it within *La mueca*: "In a world ordered by sexual imbalance, pleasure in looking has been split between active/male and passive/female. The determining male gaze projects its fantasy onto the female figure, which is styled accordingly."[26] In the film-making (and in much of the play), the looking is done by the four intruders, mainly Sueco—meaning that his gaze is the active one that shapes the action that it records. The spectators' position is a masculine one formed by Sueco. In this way, we come to the idea of a masculinity structured around power and control—two elements that are at the center of the four intruders' struggles with the couple and amongst themselves. However, the action we see is not really a film but the *making* of a film. Therefore, we are free to look wherever we choose—the spectators' gaze escapes from the confines of the filmmakers' in order to see whatever we choose onstage (or even offstage). In this way, Pavlovksy's play challenges this idea of gaze by liberating it. By showing the production rather than the finished product, the spectator is free to direct his/her eyes to different places, both using and undoing the traditional identifications of film. Here, then, there is a connection between theater and film and its manipulations. Theater, however, is shown to be more free.

Meta-theater is an important aspect of the play that continues along the same lines as that of the filmmaking and staging. It has a central role in the production of both the film and the theatrical spectacle itself. Along with comedy, Alfonso de Toro recognizes meta-theater as a central theme in Pavlovsky's work, one that underlines the essentially anti-mimetic nature of his theater.[27] For de Toro, Pavlovsky's meta-theater is a way to distance the spectator from the spectacle before him/her by exposing the puppets' strings. In *La mueca*, Pavlovsky shows the use of theater to provoke his audience to question their surroundings—a move that prevents the spectators from getting caught up in the argument and allows them to consider the situation before them. Pavlovsky's meta-theatricality reminds us of Martin Esslin's assertion that the objective of theater is to preserve the spectator's autonomy and to, thus, bring about thought: "Art, I think, being man's attempt to increase awareness of his own situation, has therefore the duty to shock people out of this."[28] Esslin here is discussing primarily the use of violence in theater. Nevertheless, in this particular situation, I believe that we can apply his theory to Pavlovsky's use of meta-theater vis-à-vis de Toro. Esslin champions the use of violence in theater to awaken and provoke to action the spectator—exactly the reason behind Pavlovsky's meta-theatricality, according to Alfonso de Toro, a tool to be used to provoke the spectator to thought. *La mueca* does just this; it provokes its spectators to think and consider the issues presented—the use of violence, the construction of identities—in order to evaluate the situation and its effects for the present and the future.

La mueca's meta-theatricality can be observed from the beginning of the first act, when the men are setting the stage for the couple's entrance and the filming that they are about to do. At this moment, the spectator observes much of their habitual preparations, one of which appears to be the use of a script in the making of the film, an idea from which they depart on this occasion:

> *Sueco*: [...] ¡Ensayemos! (*Todos van en busca de su equipo: grabador, filmadora, máquina, etcétera*).
> *Turco*: —¿Pero qué hacemos hoy? ¿Se puede ver el guión?
> *Flaco*: —Hoy no hay guión.
> *Aníbal*: —¿Cómo que no hay guión?
> Turco: —¿Y entonces qué hacemos?
> *Flaco*: —Vamos a ver cómo viene el asunto y después improvisamos...
> [*Sueco*: [...] Let's rehearse! (Everyone goes looking for his equipment: tape recorder, video recorder, typewriter, etc.).
> *Turco*: —But what are we doing today? Can we see the script?
> *Flaco*: —Today there is no script.
> *Aníbal*: —What do you mean there's no script?
> *Turco*: —And so what do we do?
> *Flaco*: —We'll see how it develops and then we'll improvise...] (Act 1: 18).

The scene, first, shows that the men's filming is a habitual one in which they normally participate with a script. Tonight, inexplicably, they decide not to follow a script and to allow the actors and the crew to react to whatever comes up. This decision seems to be an apt one for the couple since one of Helena's first comments to her husband when they enter the stage is to remark upon his acting: "¿No te das cuenta de que hacés papelones? [Don't you realize that you play the fool?]" (Act 1: 20). This observation prompts an argument over the extensive roles and performances that each one plays in their everyday lives, an opinion of Carlos and Helena that is shared by Sueco and the others. Sueco, in speaking to Carlos, remarks on the actions and roles that correspond to being a refined bourgeois man in their society—a figure deeply marked by contradiction and performance:

> Pero el problema radica en que usted no sabe hasta donde puede llegar con su sensibilidad. Usted siempre se queda a mitad de camino; su sensibilidad sólo le sirve para coquetear un rato con alguna mujer, o para comprar un buen cuadro; para leer un buen libro, para gozar de un buen espectáculo, o para saborear un buen whisky importado o tal vez en el mejor de los casos para masturbarse de vez en cuando si está muy tenso. Pero siempre en una sensibilidad pasiva, de espectador de platea cara, de burgués refinado que aprendió a mirar a los demás. A leer la historia que los demás hacen con su sangre. Pero en cambio hoy, aquí, nosotros nos sentamos en la platea y su sensibilidad se multiplicó por cien, por mil, se rompió en pedazos,

explotó en mil burbujas, y usted nos sacudió, nos violentó, nos impregnó brutalmente con otra clase de sensibilidad, fue cuando usted dejó de ser espectador y se convirtió en actor. [But the problem resides in the fact that you don't know how far you can go with your feeling. You always stop halfway; your feeling is only good for flirting with a woman, or buying a nice painting; reading a book or enjoying a nice show, or tasting a nice imported whisky or maybe in the best of cases masturbating once in a while when you're tense. But always a passive feeling, of a spectator in an expensive seat, of the refined bourgeois who watches others. Reading history that others make with blood. But in turn today, here, we are sitting in the expensive seat and your feeling has been multiplied by a hundred, by a thousand, it's shattered, exploded into a thousand bubbles, and you have shaken us up, you've made us feel awkward, you've brutally impregnated us with another type of feeling, it was when you stopped being a spectator and became an actor] (Act 2: 40).

In this scene, Sueco points out first the role that Carlos has learned to play as the "burgués refinado," one that includes certain details and attitudes regardless of individual desires. By highlighting the details that characterize Carlos' existence, Sueco questions Carlos' positioning and independence in choosing who he is and how he got there. Instead, Sueco reveals Carlos' decisions and actions as staged, as a performance that attempts to define him as any other bourgeois man at that point and time. Carlos participates in society with pre-coded behavior—he is, in fact, performing Carlos the bourgeois.[29] The awakening that the four intruders are provoking shows itself to be the first possibility of revealing who Carlos is. Similarly, the reversal in roles that Sueco recognizes in Carlos' actions suggests an increase in possibilities when the tables turn and the actors become the spectators.

The filming of Carlos' and Helena's hypocrisies continues to the very end of *La mueca* and reveals the reason behind the intruders' entrance into this couple's lives. Sueco's analysis of the roles that Carlos and Helena play persists as he details the need to record and archive their way of life before it vanishes—an explicit judgment from the play that suggests that their lives cannot continue along the route that they are following:

> Sabe usted que... bueno, que ustedes son una clase muy especial, a quienes yo personalmente respeto y admiro; porque tienen una gran sensibilidad, por el buen gusto, ¡y además por el fuerte poder adquisitivo! Son de pura raza, y además, son fundamentalmente actores sin saberlo. Son grandes actores. Generalmente un buen actor disminuye su rendimiento después de diez horas de trabajo. ¡Pero ustedes son capaces de actuar las veinticuatro horas seguidas, sin sentirlas! ¿Se da cuenta? ¡Para mí esto es sensacional! Comprende, ahora, ¿por qué los elegimos como candidatos? ¡Dónde íbamos

a encontrar tanta hipocresía y tanta podredumbre junta! Ustedes son de pura raza. El único problema es que son de una raza que se está extinguiendo poco a poco. Por eso cuando encontramos ejemplares como ustedes, tratamos de documentarnos lo mejor posible. Después los archivamos.

[You know that...well, you two are a very special type, whom I personally respect and admire; because you have great feeling, because of the good taste, and also because of your strong purchasing power! You're pure, and you're also fundamentally actors without knowing it. You're great actors. Generally a good actor lowers his performance after ten hours of work. But you're capable of working twenty-four hours in a row, without feeling it! You realize? For me, that's fantastic! You understand now, why we chose you as candidates? Where were we going to find so much hypocrisy and corruption together! You two are a pure race. The only problem is that you're part of a race that is slowly dying out. That's why when we find examples like yourselves, we try to document it as much as possible. Then we archive it] (Act 2: 45).

Rather than focus on Carlos' role-reversal uncovered in the earlier quote, Sueco underlines the daily, weekly, monthly performance in which Carlos and Helena engage in, in order to conduct the lives they've chosen for themselves. Sueco unveils the construction behind this bourgeois existence and reveals his reasons for choosing to film the two of them. In this way, their filming is carried out with the excuse that they are preserving for posterity an existence that will become obsolete. With this revelation of his analysis of the situation, Sueco conveys the reality behind Carlos' and Helena's lives—lives created from hypocrisy and hidden by an omnipresent performance that is nothing more than a façade. We are reminded of Jorge Dubatti's classification of Pavlovsky's theater as that of defining and recording the "momento histórico-social [historical-social moment]" in which it finds itself. Dubatti states that Pavlovsky's theater in the 1970s changes course and becomes more focused on the possibility of change.[30] Pavlovsky's play and characters, then, challenge representations of identity and their implications by highlighting characters' performances and the violence they use to create their selves, as we will see below.

Pavlovsky's play is structured around various acts of violence that help to advance the story and attempt to solidify the identities of the characters that he has put in motion. Perhaps the most significant violence is the constant struggle for control between, most notably, Sueco and Flaco, and the continuous physical and verbal violence among the four. This is shown at the beginning of the play when Sueco attempts to demonstrate his dominance over the other three:

Aníbal: (*Mirando afuera*).—Ahí llega el Flaco; ¿le abro? (*El SUECO no contesta*).

Turco: (*Al Sueco, con los zapatos en la mano*).—¿Los zapatos dónde los dejo?
Sueco: —¡Metételos en el culo! (*El TURCO no sabe qué hacer. Se ríe y deja los zapatos en el rincón, donde ya están los otros*). *El FLACO golpea la puerta, cada vez más* fuerte).
Sueco: —¿Y? ¿Qué te parece, Aníbal? (*Pausa*). ¿Qué te parece? (*Le toca la cabeza cariñosamente*).
Aníbal: (*Devotamente*). –Es bárbaro esto. (*Mirando a todos los lados*).
Sueco: —¡Si querés un trago servite...! (*ANÍBAL asiente*).
Turco: —¿Las valijas las dejo ahí?
Sueco: —No; subilas y dejalas arriba. (Pausa). ¡Vamos, dale! (El *TURCO* sube con las valijas. A *ANÍBAL*): ¡Abríle! (*ANÍBAL abre. Entra el FLACO. Es el mejor arreglado y más distinguido de los cuatro. Se hace claro de que pareciera responder a una clase social más elevada. El SUECO lo espera. El FLACO lo mira fijo. Se enfrentan. ANÍBAL los mira muy tenso temiendoque ocurra algo. El FLACO parece no tener miedo*).
[*Aníbal*: (*Looking out*).—Here comes Flaco; should I open to him? (*Sueco doesn't answer*).
Turco: (*to SUECO, with his shoes in his hand*). –Where should I leave my shoes?
Sueco: —Shove them up your ass! (*TURCO doesn't know what to do. He laughs and leaves his shoes in the corner, where the others are. FLACO knocks on the door, each time harder*).
Sueco: —So? What do you think, Aníbal? (*Pause*). What do you think? (*He touches his head affectionately*).
Aníbal: (*Devotedly*). –This is great. (*Looking around*).
Sueco: —If you want a drink, go ahead...! (*ANÍBAL assents*).
Turco: —Should I leave the suitcases here?
Sueco: —No; take them upstairs and leave them there. (Pause). Let's go, come on! (*Turco takes up the suitcases. To Aníbal*): Open the door! (*ANÍBAL opens it. FLACO enters. He is the best dressed and most distinguished of the four. It is made clear that he belongs to a higher social class. SUECO waits for him. FLACO stares at him. They come face to face. ANÍBAL watches them very tensely, afraid something may happen. FLACO doesn't seem to be afraid*)] (Act 1: 10).

For the most part, Sueco reigns over the men and they look to him for guidance as their leader. His manner with them is erratic; at times, he is considerate and seems to ask for their approval ("¡Si querés un trago servite...! [If you want a drink, go ahead...!]"). At others, he rules over them with an iron fist ready to crush any sign of stupidity or disloyalty ("¡Metételos en el culo! [Shove them up your ass!]"). It is in this irrationality where the most violent acts lie, given that neither Sueco's men nor the spectator can learn to predict his reactions and must decide how to avoid his mood swings. In contrast, Flaco's very image (the stage directions above indicate that he belongs to a higher social class) is one that questions Sueco's superiority among the group and he does not hesitate to test Sueco's leadership in front of the others, something that Sueco continually tries to contest or undo, as seen above. Flaco's appearance sets him apart from the other men and threatens Sueco's control. Neither man gives in to the other, preparing the men and the spectators for a struggle for control, one that will be constantly shifting back and forth:

> *Flaco*: —Querías dejarme afuera, ¿no? ¡Me quedé comprando cigarrillos, por eso me retrasé! Sos injusto, Sueco, ¿eh? ¡A veces pienso que sos tremendamente injusto...!
> *Sueco*: —¡Caminá, artista! Caminá antes que te devuelvan empaquetado a tu mamita. (*El FLACO cruza haciéndose el maricón. Mira el lugar. El Sueco parece interesado por la opinion del FLACO*). ¿Y... qué te parece? ¿Te gusta, pituquito? (*El FLACO no le contesta. Mira los zapatos. Están los tres pares en un rincón*).
> *Flaco*: —¿Van a venir los Reyes Magos? (*Ríen los tres mientras el FLACO se saca los zapatos.*
> [*Flaco*: —You wanted to lock me out, no? I was buying cigarettes, that's why I was late! You're not fair, Sueco, eh? Sometimes I think that you're tremendously unfair...!
> *Sueco*: —Walk on, artist! Walk on before you're sent packing back to your mommy. (*FLACO crosses the stage playing the fag. He looks at the place. SUECO seems interested in FLACO's opinion*). So...what do you think? Do you like it, honey? (*FLACO doesn't answer. He looks at the shoes. The three pairs are in the corner*).
> *Flaco*: —Are the Three Magi coming? (*The three of them laugh while FLACO takes off his shoes*)] (Act 1: 10–11).

While Sueco seems to occupy the role of leader, Flaco constantly proves that he is outside the control of any other person, a fact that can be attributed to his belonging to a higher social class than the others. Turco and Aníbal, in turn, show very little resistance to the leadership of one of the other men, never challenging outright their role as followers. Sueco appears to be their leader given that he orders them to do things, using the threat of violence against them. He is the one who leads through fear and violence rather than respect and admiration, two qualities that Flaco seems to be able to inspire in Turco and Aníbal. Nevertheless, Flaco and Sueco also look to one another for approval, as Sueco does when Flaco appraises the apartment, thus unbalancing further their struggle and the power of the group.

Sueco's use of physical violence as a leadership tool is shown soon after Flaco's entrance into the apartment, when he begins to question Turco's actions. Both his attitude and gestures toward Turco and his words make evident the thin line he walks between self-control and abandon, a tactic that he uses in order to dominate his other gang members. It is this fine line which reveals the purpose of violence in this play:

> *Sueco*: —¿Dónde dejaste las valijas?
> *Turco*: —En la bañadera.
> *Sueco*: —No jodas, ¿dónde dejaste las valijas? (*Lo agarra de la remera y lo está ahorcando casi*).
> *Flaco*: —Largalo, Sueco; no te pongas grosero... Se te van mal cuando te ponés grosero.
> *Sueco*: (*Al TURCO*). –Te pregunto dónde dejaste las valijas. (*Le pega un bife*).
> *Turco*: —Ya te dije, Sueco; en la bañadera.
> *Aníbal*: (*Va al TURCO y lo libera tomándolo él*). —¿Estaba llena?
> *Turco*: —Si estaba llena ¿qué?
> *Aníbal*: —La bañadera.
> *Turco*: —¿De qué?
> *Aníbal*: ¡De agua, boludo!
> *Turco*: No; estaba vacía. (*El SUECO mira el diálogo asombrado, como pensando, y estalla*):
> *Sueco*: Por favor, no dialoguen, ¡eh! No dialoguen ustedes dos sin pedir permiso porque son capaces de enloquecer a un burro. (*Lo agarra de una oreja al TURCO*). Escuchame, Turco; vení contame, ¿cómo fue que se te ocurrió dejar las valijas en la bañadera? Vení, querido; vení, sentate.
> *Turco*: —No sé, Sueco, no sé. ¡Abrí las puertas de las piezas y no me animé a entrar! Nada más...

> [*Sueco*: —Where did you put the suitcases?
> *Turco*: —In the bathtub.
> *Sueco*: —Don't fuck around, where did you put the suitcases? (*He grabs him by the t-shirt and almost chokes him*).
> *Flaco*: —Cut it out, Sueco; don't be crude... Things go badly when you're crude.
> *Sueco*: (*To Turco*). –I'm asking you where you put the suitcases. (*He slaps him*).
> *Turco*: —I told you, Sueco; in the bathtub.
> *Aníbal*: (*He goes to Turco and frees him by taking a hold of him*). –Was it full?
> *Turco*: —Was what full?
> *Aníbal*: —The bathtub.
> *Turco*: —Of what?
> *Aníbal*: Of water, stupid!
> *Turco*: No; it was empty. (*Sueco watches the dialogue with surprise, as if thinking, and he explodes*):
> *Sueco*: Please, don't talk, eh! Don't you two talk without asking for permission because you're capable of driving a donkey crazy. (*He grabs Turco's ear*). Listen to me, Turco; come, tell me, how did it occur to you to put the suitcases in the bathtub? Come, dear, come, sit down.

Turco: —I don't know, Sueco, I don't know. I opened the doors to the rooms and I didn't feel like going in! That's it...] (Act 1: 11).

In this scene, Sueco attempts to gain control over the group through his physical and mental control over Turco. He tries to exert a physical power over a member of the group whose actions have stepped out of line. Sueco must maintain control of both his manner of questioning and of his men, especially when the latter attempt to think for themselves, as we see when Aníbal interrupts the two to find out more about Turco's actions. In this exchange, the spectator can see the introduction of class differences into the gang members' milieu when Flaco accuses Sueco of acting "grosero [crude]," an insult that is meant to stop him and, in a subtle way, question his authority over the group. This questioning continues when Flaco decides to interpret Turco's actions as imaginative and revolutionary:

Flaco: —¿No te parece una maravilla, Sueco? ¡El Turquito tuvo miedo de poner las valijas en los cuartos y lo confiesa! ¡Tuvo miedo!...y entonces resolvió por su propio libre albedrío dejar las valijas dentro de la bañadera. ¡Es magnífico! ¡Un acto de pureza! (*El Sueco lo mira y el Turco parece muy contento de la explicación del FLACO*). Lo que pasa es que el Turquito tiene imaginación, mucha imaginación. Toda la lógica indica que las bañaderas no son buen lugar para dejar valijas, pero el Turco es diferente, ¡siempre fue diferente el Turco! Es un hombre con imaginación; ¡y desafiando las leyes del sentido común y de la lógica aristotélica, nuestro buen amigo, nuestro compañero del alma, es capaz de despojarse de todos los prejuicios de la educación tradicional y en un verdadero acto revolucionario, en un verdadero orgasmo imaginativo, sorprender el mundo con una actitud aparentemente insólita dejando las dos valijas dentro de la bañadera! Imaginación que traduce la inquietud de una época convulsionada. Sí, señores...de la imaginación al poder. (*Aplausos y vivas del TURCO y ANÍBAL*).
Turco: —¡Eso! De la imaginación al poder.
Sueco: (*Pegándole una tremenda patada en el trasero*). —¡Imaginación sí, pero no joder!
Aníbal: (*Con una copa en la mano*). –Qué cultura, viejo; qué cultura. ¿Viste cómo habla el Flaco, Sueco? ¡Parece que le brotaron las palabras.

[*Flaco*: —Don't you think it's marvelous, Sueco? Turquito was afraid to put the suitcases in the rooms and he admits it! He was afraid!...and so he resolved with his free will to put the suitcases in the bathtub. It's magnificent! An act of purity! (*Sueco looks at him and Turco seems happy with FLACO's explanation*). What happens is that Turquito has imagination, a lot of imagination. All logic indicates that bathtubs are not a good place to put suitcases, but Turco is different, Turco was always different! He's a man with imagination; and by challenging the laws of common

> sense and Aristotelian logic, our good friend, our companion of the heart, is capable of giving up all the prejudices of traditional education and in a truly revolutionary act, in a truly imaginative orgasm, surprise the world with an apparently unusual attitude of putting the suitcases in the bathtub! Imagination that translates the anxiety of a convoluted time. Yes, gentleman, from the imagination to power. (*Applause and vivas from Turco and Aníbal*).
> *Turco*: —That's it! From the imagination to power.
> *Sueco*: (*Kicking him hard in the behind*). —Imagination yes, but don't fuck around!
> *Aníbal*: (*With a glass in his hand*). –What learning, man; what learning. Did you see how Flaco talks, Sueco? It's like the words flow from him] (Act 1: 11–12).

Flaco's interruption here serves to challenge Sueco's right to lead the group and shows that Flaco is the one who inspires respect and admiration in the men. Embodied in these two men, two ways of gaining and retaining power are illustrated. Sueco's method of leadership is to bully and impose his way on the others. Flaco's words and actions, in turn, advocate a questioning of the status quo and a use of innovative thinking in order to re-consider the world—a philosophy that was popular in the late 1960s and draws inspiration from the free thinking characterized by the time.[31] Nevertheless, implicit in Flaco's defense of Turco's actions is an irony that scoffs at both Turco and the situation. Flaco is a natural leader who doesn't need to intimidate or to exert physical violence over the others to lead. He uses his wit to control and rule the others while he simultaneously mocks them. He defends Turco's unusual decision to put the suitcases in the bathtub by lauding his "imagination," implicitly underlining the absurdity of Turco's actions and his defense, destabilizing both the scene and the audience's expectations of the play. By inspiring laughter, Flaco reminds us of the ability Mikhail Bakhtin identified in laughter to change images:

> Laughter has the remarkable power of making an object come up close, of drawing it into a zone of crude contact where one can finger it familiarly on all sides, turn it upside down, inside out, peer at it from above and below, break open its external shell, look into its center, doubt it, take it apart, dismember it, lay it bare and expose it, examine it freely and experiment with it. Laughter demolishes fear and piety before an object, before a world, making of it an object of familiar contact and thus clearing the ground for an absolutely free investigation of it.[32]

By making fun of Turco's acts, Flaco wrests control and power from Sueco and instead takes it himself. His new interpretation allows the men to question and resist Sueco's version and, instead, gives him the power to re-evaluate and explain the situations according to a different plan of action. His ability to inspire laughter affords him a control that Sueco is never able to manage.

Laughter takes on an interesting role in *La mueca* in connection to violence. The first, most obvious appearance of the use of laughter as a form of violence is found in the first act in reference to Turco, the character who shows himself to be the least intelligent of the four. While staging a sword fight between Turco and Aníbal, with Flaco watching, Flaco and Aníbal begin to mock Turco and his alleged Cossack ancestors:

> Aníbal: —Cosaco de las pelotas. (*Le tira un sablazo en la zona baja*). ¡Turco de mierda! (*El TURCO esquiva el golpe de un salto acrobático*).
> Flaco: —Los cosacos eran rusos, animal. (*Riéndose a carcajadas*). Y vos sos turco, Turquito. ¿De dónde vas a tener un abuelo cosaco, vos?
> Turco: (*Enojado*). —¡Te digo carajo que mi abuelo era cosaco! (*Bajando el arma*). Me lo dijo el viejo, entendés, y yo al viejo le creo.
> Aníbal: —Subí el sable que te saco la cabeza. No hagas política, Turco, te falta información.
> Flaco: (*Riéndose a carcajadas*). –Está bien, Turquito, no te enojes. (*Se tienta cada vez más*).
> Turco: (*Cada vez más engranado*). —¡No te rías de mi abuelo, eh! Te lo digo en serio, carajo. ¡No me gusta que te rías de mi familia! (*El FLACO se ríe hasta el paroxismo*).
> Flaco: —Pero, Turquito... Te fifás a tu hermana, ¿y ahora salís defendiendo el escudo de la familia? No sigas hablando que me muero de risa.
> Turco: —Mirá, Flaco de mierda que... (*Amaga sobre él. El FLACO se cae del otro lado del sillón, se oyen estertores de risa. Aníbal provoca al Turco desde otro lado*).

> [Aníbal: —Cossack, my ass. (*He strikes a blow with the sable in the lower zone*). Shitty Turk! (*Turco evades the blow with an acrobatic jump*).
> Flaco: —The Cossacks are Russians, animal. (*Laughing heartily*). And you're a Turk, Turquito. Where are you going to have a Cossack grandfather from? You?
> Turco: (*Angry*). —I'm fucking telling you that my grandfather was a Cossack! (*Lowering his weapon*). My old man told me, you see, and I believe my old man.
> Aníbal: —Raise your weapon or I'll cut off your head. Quit the BS, Turco, you lack information.
> Flaco: (*Laughing heartily*). –It's okay, Turquito, don't get mad. (*He gets tempted more and more*).
> Turco: (*More and more angry*). –Don't laugh at my grandfather, eh! I'm serious, damn it. I don't like you laughing at my family! (*FLACO laughs himself into a stupor*).

Flaco: —But, Turquito...You screw your sister, ¿and now you come out defending your family shield? Don't talk anymore, I'm dying of laughter.
Turco: —Look, Flaco, you shit...(*He makes as if to hit him. FLACO falls down the other side of the chair, rattles of laughter are heard. Aníbal provokes Turco from the other side*)] (Act 1: 14).

In this scene, laughter is used against Turco by his friends to demean him and that which he holds dear. By laughing at his claims to a Cossack ancestry, Flaco and Aníbal question his identity and familial honor and, thus, render him less than what he himself believes to be. Mockery, then, is used here as a way to break down one's beliefs and cause one to doubt. Here it is used to mock and destroy, an important aspect of laughter as identified by Bakhtin: "it is a view of the world in which all important value resides in openness and incompletion. It usually involves mockery of all serious, 'closed' attitudes about the world, and it also celebrates 'discrowning,' that is, inverting top and bottom in any given structure."[33] Laughter is what is used to turn a situation on its head and cut something down to size, as Turco finds out when Flaco uses his wit against him.

Laughter is an important element of *La mueca*. Alfonso de Toro examines how Pavlovsky's theater uses comedy as a way to capture his audience while simultaneously revealing more about his characters: "ésta se descubre por un lado como un medio de cautivar a un público, de seducirlo a través de una especie de obra costumbrista barata, campechana; por otro, la comicidad se revela dentro del contexto de la vida de los personajes y de sus acciones como un arma del cinismo [this is seen on one hand as a way to captivate the audience, to seduce them through a type of cheap, good-natured play of customs, on the other hand, the comedy reveals itself within a context of the characters' lives and of their actions as a weapon of cynicism]."[34] In the above scene, we can detect Pavlovsky's double objective, in that the tension between the men captures the audience's attention and alerts them to a possible conflict. At the same time, the scene is created so that we connect with Flaco—the suave, fearless one—against Turco who is portrayed as simple-minded and easily controlled. Nevertheless, de Toro's second function of laughter in Pavlovsky—cynicism—is, in my view, more important in reference to this scene. Cynicism is a common reason for laughter; it is used when we believe that innocence has vanished, and it can be used as a tool against others to mock their naiveté. Flaco, accompanied by Aníbal, reveals his own cynicism toward Turco's efforts to bolster his self-image. Flaco's cynical laugh undoes Turco's efforts and exposes him as a fraud. When he answers "Está bien, Turquito, no te enojes [It's okay, Turquito, don't get mad]," he deepens the wound by pretending to yield, while his use of the diminutive with Turco's name stresses and increases the unbridgeable difference of class and intellect between the two men. Similarly, the audience begins to pity Turco for not being as urbane as Flaco, for believing his aggrandizing father, for wishing he were something else. Flaco's laughter here violently undoes Turco and his desires and leaves him with less than he had before, making the audience skeptical of the characters and their assertions. Similarly, the situation unmasks Flaco and his own prejudices. He is revealed to not be like the object of his laughter—he is something else, confirmed by his appearance and way of speaking. Flaco is able to use laughter against the others but curiously remains above the mocking, biting power of humor that is manifested in the scenes quoted above.

Toward the end of the first act, the spectator encounters another moment where laughter becomes a violent tool that distorts the action around it. At this point, though, the laughter is a different sort of instrument given that it is not inspired by another's actions and its purpose is not to undo another character. Instead, given its extreme manifestation in this scene, laughter is the manifestation of a violent undoing. This comes about, thanks in part to a drink that the intruders give to Carlos and Helena, when Carlos begins to laugh hysterically, causing alarm in the four men:

> *Sueco*: —¿En serio no se te fue la mano con el lisérgico?
> *Flaco*: (*Asustado*). —¿Cuánto le pusiste?
> *Carlos*: (*Riéndose a carcajadas*). —¡Lisérgico...! !La gran puta!
> *Turco*: —Le puse lo de siempre.
> *Sueco*: —Bueno, será muy sensible el gentleman. Déjenlo que se desahogue. Le va a hacer bien. (*CARLOS se tira en el sillón y retuerce de risa. La escena es grotesca*).
> [*Sueco*: —Seriously don't you think you went overboard with the LSD?
> *Flaco*: (*Afraid*). –How much did you give him?
> *Carlos*: (*Riéndose a carcajadas*). –LSD...! Shit!
> *Turco*: —I gave him the same as always.
> *Sueco*: —Well, the gentleman must be very sensitive. Leave him alone to get over it. It'll do him good. (*CARLOS throws himself onto the chair and twists with laughter. The scene is grotesque*)] (Act 1: 24).

Laughter in this scene is destabilizing: for Carlos, he is overcome by laughter and unable to do anything more than laugh, the intruders are taken by surprise by the uncharacteristic reaction to their normal action. Moreover, the violence of Carlos' laughter is important and cuts through the act itself to create a new experience for all involved. Rather than being used as a weapon to cut down the object of the laughter as we saw earlier with Turco, Carlos' laughter undermines the action of the four interlopers and interrupts the action on stage, despite the fact that it appears to be an unconscious act on his part. In this way, his laughter is a violent break in the action—it is not what was expected of him, and the four intruders don't know how to react.

Laugher becomes a rupture with the predictable and advances the action of the play while it simultaneously stops the characters' movements and provokes self-questioning. If we remember Alfonso de Toro's double function of laughter, Carlos' hysterical laughing does captivate both his audience (the intruders) and the play's audience. Nevertheless, rather than demonstrate or provoke cynicism, the laughter interrupts theatrical cynicism in that it cuts short what is expected. Whereas cynicism is inspired by routine and stagnation, Carlos' hysterical laughter breaks routine and prevents either the intruders' or the audience's cynicism by being unpredictable. It directly engages the audience through surprise and, thus, recalls Bakhtin's belief that carnival does not recognize a distinction between actor and spectator.[35] Pavlovsky wishes to stir Carlos' audiences and provoke them to question the situation before them. In this way, laughter awakens its public from cynicism and inspires original action. Thus, laughter in *La mueca* is a weapon that is used against the characters and to engage with the audience.

Social class is registered in this laughter and is an important issue that separates the characters and initiates the action in *La mueca*. The four men that enter into the apartment seem to belong, at first glance, to the working class (although Flaco clearly distinguishes himself from the others in both dress and speech). Nonetheless, their actions (entering an apartment that is not their own, filming a couple engaging in sexual activity) set them apart from the couple, who, in turn, belong to the bourgeoisie, as evident in their home and actions. Class difference introduces the inevitable clashes and offenses that take center stage throughout much of *La mueca*. The fact that the bourgeois couple has had very little interaction with others and cannot understand their way of life is made evident by the constant offenses and misunderstandings that take place between the two factions. One of the most significant of these offenses Pavlovsky places at the beginning of the second act. Not being able to understand the intruders' actions in any other way, Carlos asks if they are communists—a question that ignites a fiery reaction:

> *Sueco*: ¡Hablá o te mato! ¿Qué dijiste de nosotros?
> *Carlos*: (*De golpe*). Le pregunté al señor si eran comunistas.
> *Sueco* (*Soltándolo bruscamente*). ¡Ooohhh! (*Agarrándose la cabeza*). ¿Dijo eso...? (*Casi llorando*). ¡Insensato! ¡Nunca debió decir esto! ¡Nunca! (*Se tapa la boca con la mano y se la muerde. Acá la escena no se debe saber si*

es real o es fingida, tan dramática es para el Sueco la palabra "comunista"). Eso es una confusión atroz. Nos hiere. ¡No merecemos esto! ¡No lo merecemos! (Llora).
Carlos: Permítame, señor, le hice una pregunta, ¡no quise ofenderlos!
Sueco: ¡Somos artistas! (*Cada palabra es pronunciada como si fuera definitiva. Pone la cara al lado de CARLOS. Lo toma del cuello*). Somos artistas y no comunistas. ¡Recuérdelo!
[*Sueco*: Talk or I'll kill you! What did you say about us?
Carlos: (*All at once*). I asked the gentleman if you were communists.
Sueco (*Brusquely letting him go*). Ooohhh! (*Grabbing his head*). ¿You said that…? (*Almost crying*). Insensitive! You should never say that! Never! (*He covers his mouth with his hand and bites it. Here one shouldn't be able to tell if the scene is real or fake, that's how dramatic the word "communist" is for Sueco*). That's an atrocious confusion. It hurts us. We don't deserve this! We don't deserve it! (*He cries*).
Carlos: Excuse me, sir, I asked him a question, I didn't want to offend you!
Sueco: We are artists! (*Each word is pronounced as if it were definitive. He puts his face next to Carlos. He takes him by the neck*). We are artists and not communists. Remember it!] (Act 2: 32).

Whether Sueco's anger at being called a communist is real or false, this scene demonstrates the distance that exists between Carlos and Sueco and Carlos' inability to understand the intruders' lives. The two come from different worlds that are unbridgeable at this moment. Furthermore, Sueco's reaction is meant to provoke laughter in the audience given that the violence of his response is unexpected and surprising. He inspires laughter that questions and degrades Carlos in front of his wife and the spectators.

Sueco's reaction to being called a communist is particularly interesting when we consider the role that communism and the arts played in the 1960s and 1970s in Latin America. As stated in the introduction, this time period was marked by revolution and upheaval, both close to home (Cuba) and farther afield (Paris, Vietnam). As Claudia Gilman describes in *Entre la pluma y el fusil: debates y dilemas del escritor revolucionario en América Latina*, the connection between the arts and political engagement was particularly strong during the 1960s and the early 1970s, and this generally meant a connection between the figure of the intellectual and the left.[36] As the decade progressed, this meant a connection to the word "revolution" and, with it, violence: "Para la izquierda, a medida que avanzaban los años, la noción de revolución iba a llenar toda la capacidad semántica de la palabra 'política'; revolución iba a ser sinónimo de lucha armada y violencia revolucionaria [For the Left, as the years advanced, the idea of revolution would fill all the semantic capacity of the word 'politics;' revolution would be a synonym of armed struggle and revolutionary violence]."[37] In this way, when Carlos asks

Sueco if he is a communist, he is referencing a geographical concept of violence for political means, an idea that also connected to the Argentine reality of armed struggle that defined the post-*cordobazo* world.

In Argentina, as detailed earlier, the political and social contexts were defined by the unrest and economic stability of the moment. Guerrilla groups on both the left and the right clandestinely practiced kidnappings and assassinations in an attempt to grab power and financing. Some of these left-wing secret organizations espoused a radical form of Peronism, such as the *Montoneros*, that hoped to advance a revolution such as the Cuban one. The ERP (*Ejército Revolucionario del Pueblo*), on the other hand, went much further in their radicalization in that they wanted to "transcend" the state and spark a pan-American revolution against imperialism.[38] By connecting the violent intrusion of the four men into his house, Carlos points to these political acts and paints a picture in which all transgressions can be traced back to a political idea. Sueco, however, rejects this connection that would interpret his group's actions as contributing or advancing these contemporary ideas. Instead, Sueco and the other men become autonomous artists whose art cannot be easily defined or labeled.

This autonomy is also consistent to a certain degree with the Latin American revolutionary artist in that Claudia Gilman recognizes his/her refusal to subordinate his/her views to the Communist Party, particularly in the form of Soviet Stalinism.[39] This independence can be seen even more sharply in the scene above. The intruders refuse to adhere to a conventional way of thinking and acting. Instead, their actions are only accountable to their own rationale of the world. They are truly independent and this position is what allows them to evaluate Carlos and Helena.

Here, Carlos is unable to comprehend the position of the intruders given that they do not adhere to his labeled way of thinking. However, Carlos is not the only one who cannot grasp the four men's situation; Helena also misjudges the circumstances in which she finds herself and attempts to divide the men along class lines—the characteristic that she believes to be most important in defining who these men are. Simultaneously, she collapses social class with authority, believing that Sueco's membership in a lower social class would not permit a space of leadership. In this quote, she sees that Sueco is the leader and questions this by instead placing Flaco in that space:[40]

> *Helena*: ¿Así que para subir las escaleras le tiene que pedir permiso a éste?
> *Flaco*: ¡Cállese señora, que le conviene, por favor!
> *Helena*: No entiendo qué hace usted aquí. Usted es sapo de otro pozo, ¿no le parece?

Sueco: (*Ríe*). ¡Te calaron, Flaco!... (*El FLACO, que se iba a ir por las escaleras, se vuelve y se enfrenta con la SEÑORA*).
Flaco: Mirá, nena. Si lo que pretendés es meter cizaña entre nosotros te prevengo que los cachetazos te los voy a dar yo, ¿me entendés? La ironía y la inteligencia metételas en el culo y no confundás cortesía con boludez. Mirá que yo conozco muy bien a las minas como vos... ¿eh?
Helena: No ves que hablamos el mismo idioma... ¡Sos un sapo de otro pozo!
[*Helena*: So, you have to ask this one permission to go upstairs?
Flaco: Be quiet, ma'am, for your own good, please!
Helena: I don't understand what you're doing here. You're a horse of a different color, don't you think?
Sueco: (*Laughs*). They've seen through you, Flaco!... (*FLACO, who was going to go upstairs, turns around and confronts the WOMAN*).
Flaco: Listen, chick. If what you're trying to do is sow discord among us, I warn you that I'll be the one to throw you the punches, got it? You can shove irony and intelligence up your ass and don't confuse courtesy with stupidity. Because I know what kind of girl you are... eh?
Helena: You don't see that we speak the same language... You're a horse of another color!] (Act 2: 29).

For Helena, Flaco is like Carlos and herself, and she cannot fathom why he chooses to identify with the others rather than with those of his "own kind." Nevertheless, when she tries to exploit their similarities, she is immediately spurned, although she does not give up. Flaco, in turn, changes his attitude toward her in an attempt to break the link between the couple and himself. Rather than warn her respectfully of the consequences of her actions (all of which he did earlier, including referring to her as *señora* and with the formal *usted* form), he modifies his speech, using the informal Argentine *vos* and local slang and threatening her with physical violence—speech patterns that attempt to distance him from his bourgeois ties and bring him closer to his companions.

Throughout *La mueca*, Pavlovsky uses the violent gestures of one character against another as a tool to understand the construction of identity and self in both the play and the surrounding community. The spectator witnesses constant bickering and arguing between the intruders and the couple and amongst these groups. Nevertheless, physical violence is used to coerce a reaction from one character or to force another to take a certain action. Violence is, similarly, what is used to construct gender in *La mueca*, permitting some characters a space of masculinity while others are relegated elsewhere, as seen in the scenes analyzed below. As Judith Butler has stated, gender (not to be confused with sex) is performative: "gender reality is performative which means, quite simply, that it is real only to the extent that it is performed."[41] This means that one's actions and performance define one's gendered space—gender must be constantly performed against a fictional ideal. Performing dominance is masculine while the opposite is a feminine space, regardless of sex. Performance in *La mueca* is conducted through violence against the other in order to subjugate and control. As seen in the following scene where Helena tries to force her husband to stand up to the intruders, violent gestures are used by the men against the couple to divide them and to, then, bring about the desired reaction for their filming:

> *Helena*: Demostrale que no le tenés miedo. ¿Por qué una vez en tu vida no te comportás como un hombre?
> *Sueco*: ¿Qué fue lo que dijo arriba?
> *Helena*: ¡No le hablés que después es peor!
> *Sueco*: Hablá que te conviene, infeliz. (*CARLOS está con la cabeza baja llorando impotente. El FLACO filma de todos los ángulos. Tomándolo del pelo y levántandole la cabeza*). ¡Hablá maricón!
> *Helena*: (*Se acerca al Sueco*). ¡No hablés, Carlos! (*El SUECO le pega un bife a CARLOS con una mano mientras con la otra lo tiene amarrado del pelo. HELENA se abalanza sobre el SUECO, el SUECO le pega un golpe. HELENA cae sobre el sillón. El golpe fue en la cara, bastante fuerte. CARLOS no se ha movido ni intentado defensa alguna. El FLACO, el TURCO y ANÍBAL están en plena tarea, totalmente ausentes, como si fuera una escena artística pura*).
>
> [*Helena*: Show him you're not afraid of him. For once in your life why don't you be a man?
> *Sueco*: What was it you said upstairs?
> *Helena*: Don't talk or it'll be worse later!
> *Sueco*: Talk, for your own good, you wretch. (*Carlos has his head down, crying impotently. Flaco is filming from all angles. Taking him by the hair and lifting up his head*). Talk, you fag!
> *Helena*: (*She approaches SUECO*). Don't talk, Carlos! (*Sueco slaps Carlos with one hand while with the other he holds him by his hair. Helena pounces on Sueco, Sueco strikes her. Helena falls on the chair. The rather strong blow was to her face. CARLOS hasn't moved or tried any defense. FLACO, Turco and Aníbal are absorbed by their tasks, totally absent, as if it were a purely artistic scene*)] (Act 2: 31).

Sueco's physical beating of both Carlos and Helena in this scene is used to show the strength of one man over another, emasculating the latter and placing him in a space of inferiority—proving that one is a gendered male (meaning that he can dominate the others) and the other is not. Helena goads her husband into proving to these men and to her that he is a man, meaning that he will not bend to their will despite physical violence, that he will not be dominated. Being a man means not being afraid and not bending to another. It is interesting to point out that in this scene, and in the play in general, though Carlos demonstrates moments of standing up for himself, Helena is the one to attack the intruders, both physically and verbally—she takes on the male gendered role at this point. Nevertheless, here we see that Sueco dominates the scene, towering over Carlos and making the latter the victim of his own requests. Helena, in turn, challenges both Sueco and Carlos and edges them both into action against her. Throughout all of these violent acts, the other three men remain on the outside, filming constantly the scene that they have helped create, as can be seen in the stage directions that end the above scene. For them, this is part of a movie that they are creating and the violent gestures of their leader are just a necessary inclusion—after all, it is simply an "escena artística pura [pure artistic scene]."

The stage directions and the gestures that these indicate play a central role in the creation of gender lines, particularly in this scene, given the way they create the characters and their identity. The actions and gestures that are delineated contribute to the construction of who each character is and how his/her identity fits into the larger picture. Moreover, the interruption of the written text provides a violent intrusion that mirrors the actions they dictate. For the reader, the italicized text jumps off the page at the reader and interrupts the dialogue as the physical actions in the representation similarly intrude upon the actors' words. In this way, the representation and the written script come together to create a similar visual effect.

Returning to the construction of masculinity, in its manifestation above, one of the men has to dominate the other in order to be considered the "man," an identity that is claimed by Sueco in the previous scene. Nonetheless, later in the same act the tables turn and Carlos begins to attack Sueco verbally and physically, rising to the challenge of defending his wife and his home. In this scene, Carlos seems to lose all control over his actions and tries to regain the command of his own house. What is interesting in what follows is how Sueco responds to this challenge and what new violent actions and reactions this response means for both Helena and Carlos:

> *Carlos*: [...] ¡Yo les voy a demostrar que sé defender lo mío hasta el final! ¡Ahora soy yo el que grita! ¿Qué mierda quieren? ¿De dónde salen ustedes, eh? ¡La guaranguería se las voy a meter en el culo! (*Al Sueco*). ¡Y vos, maricón, degenerado! ¿Qué te creés? ¿Que sos capaz de salvar a la humanidad con tus discursitos? (*Se acerca al Sueco y lo agarra*). ¡A vos te digo! ¡Yo laburé en mi vida, a quien querés joder! (*Lo empuja*). Nadie se mueva, ¿eh? ¡Cagones! Yo sé lo que es laburar, y ustedes son una manga de parásitos que habría que fusilar! Yo les cortaría el pelo y los colgaría de las pelotas en Plaza de Mayo! ¿Y vos, Suequito, quién te creés que sos? Pero, ¿a quién pretendés salvar, infeliz de mierda? ¡A ver, quién de los

tres lo defiende! ¡Miren cómo le pego al Jefe, miren! (*Le pega bifes en la cara al Sueco*).
Helena: ¡No ves que son una manga de cabrones!
Carlos: ¡Callate, Helena!
Helena: ¿Por qué me voy a callar?
Carlos: ¡Vamos, salvador de la humanidad! ¡Mesías del proletariado! ¡Defendé tu arte y tus discursos! ¡Maricón! (*Le pega otro bife al Sueco*).
Flaco: (*Muy tranquilamente*). La va a pasar peor, Carlos. ¡Mejor es que se calme!
[*Carlos*: [...] I'm going to show you I know how to defend what's mine to the end! Now I'm the one yelling! What the fuck do you want? Where'd you come from, eh? I'm going to shove your bad behavior up your ass! (*To the Sueco*). And you, fag, degenerate! Who do you think you are? That you're capable of saving humanity with your little speeches? (*He approaches Sueco and grabs him*). I'm telling you! I worked all my life, who are you trying to fuck! (*He pushed him*). Nobody move, eh? Cowards! I know what it is to work, and you're a bunch of parasites that should be shot! I would cut your hair and hang you by your balls in the Plaza de Mayo! And you, Suequito, who do you think you are? Who do you think you're saving, you shitty wretch? Let's see, which of the three will defend him! Look at how I'm hitting the boss, look! (*He slaps Sueco*).
Helena: Don't you see that they're a bunch of assholes!
Carlos: Shut up, Helena!
Helena: Why should I shut up?
Carlos: Come on, savior of humanity! Messiah of the proletariat! Defend your art and your speeches! Fag! (*He slaps Sueco again*).
Flaco: (*Very calmly*). It's going to be worse for you, Carlos. It'd be better if you calm down!] (Act 2: 34–35).

Carlos' outburst against Sueco can be understood in the context of the trauma and pain that he and his wife have suffered from Sueco's blows and the intruders' appearance in their lives. Carlos finds himself at the mercy of Sueco and his men and lashes out at their leader, thus changing for a brief moment the roles of who is in charge and who is a victim. He promises the men the same violent and humiliating treatment that he has received at their hands, graphically vocalizing the physical pain that he will give them. Nonetheless, Flaco's warning that this will only make his situation harder in the end reminds the spectator that the ones who are pulling the strings here are not Carlos and Helena. Carlos' attack, then, is a futile measure that only reinforces his own helplessness and lack of masculinity in front of the total command and "manliness" that Sueco manifests. Helena is also revealed as a helpless instigator who is not even validated by her husband.

In *La mueca*, control and violence intermingle with sexual identity and construction. As can be seen in the above scenes, Helena (and others) construct her husband's (and that of the other four men) masculine identity along the lines of domination. When Sueco is bullying Carlos for information, Helena tells her husband to be a man. As seen below, identity is formed by sexual definitions. Helena confronts Sueco about his sexual inclinations, attempting to place him in a certain fixed identity that he refuses to take on:

> *Helena*: Dígame, ¿usted es marica?
> *Sueco (Sorprendido)*: No sé lo que usted define como marica, señora. Si se refiere a tener relaciones homosexuales, le diré que a veces...sí, que a veces he tenido. Pero me parece, señora que marica en el sentido que usted lo dice...no. En ese sentido no lo soy, señora. Quede tranquila; ¡en su casa todavía no ha entrado ningún marica!
> *Helena*: Muchas gracias. Y dígame...¿qué diferencia hay?
> *Sueco*: ¿Diferencia de qué?
> *Helena (Interesada)*: ¡Usted dice que no es lo mismo tener relaciones homosexuales que ser marica! ¡Si me explica se lo voy a agradecer...!
> *Sueco*: Bueno, ese en realidad es un problema...! ¡casi filosófico...! (*Se sienta en un sillón*). En realidad un homosexual es alguien que está buscando algo y...un marica es alguien que ya lo encontró hace rato.
> *Helena*: Usted tiene alma de pedagogo, ¿no?
> [*Helena*: Tell me, are you a fag?
> *Sueco (Surprised)*: I don't know what you define as a fag, ma'am. If you refer to having homosexual relations, I would say...yes, sometimes I have. But it seems to me, ma'am, fag in the way you mean it...no. In that way, I am not, ma'am. Relax, no fag has yet entered your house!
> *Helena*: Thank you very much. And tell me...what's the difference?
> *Sueco*: What difference?
> *Helena (Interested)*: You say that it is not the same to have homosexual relations as to be a fag! If you would explain that to me, I would be grateful...!
> *Sueco*: Well, in reality that's an almost...philosophical question...! (*He sits down in the chair*). In reality a homosexual is someone who's looking for something and...a fag is someone who's found it a while ago.
> *Helena*: You have the soul of a pedagogue, no?] (Act 2: 30).

By referring to Sueco as a *marica*, Helena hopes to shame and insult him, not realizing that he is beyond her categories. Instead, Sueco changes these definitions, though he still uses classifications. He differentiates between the act and the identity, creating a division between the two for which Helena's classifications do not allow. Similarly, while Helena's insult is meant to demean him, it likewise allows for and explains Sueco's existence for her. As Judith Butler explains in *Excitable Speech: A Politics of the Performative*, "by being called a name, one is also, paradoxically, given a certain possibility for social existence, initiated into a temporal life of language that exceeds the prior purposes that animate that call. Thus the injurious address may appear to fix or paralyze the one it hails, but it also may produce an unexpected and enabling response."[42] Sueco takes on the insult and twists Helena's meaning, thus placing her in a space of inferiority and reversing the roles: he is now the cultured teacher, shining light on her ignorance. Furthermore, by appropriating the definition and differentiating between the act and the identity, Sueco's words construct his own identity. He is transforming who he is and what that means. At this point, identity becomes a choice and not an imposition of roles from the outside.

Flaco's earlier prediction that Carlos will suffer more for his outburst holds true a little later when the four intruders decide to extract a confession from Carlos as punishment. Sueco's reaction to Carlos' attack is to enter into convulsions, which makes Carlos uncomfortable with his role as aggressor and allows yet another reversal of roles. From the floor, still convulsing, Sueco yells, "¡Quiero una confesión! [I want a confession!]" and so begins the extraction of one of Carlos' undesirable acts (Act 2: 37). The scene changes quickly as the other three intruders restrain Carlos and Helena: "*CARLOS es agarrado entre los tres y atado a una silla. HELENA también es atada a una silla, y le ponen un pañuelo en la boca. La actitud ya no es violenta; parece convertirse todo en un acto de sumisión y de entrega total. Todo es preparado con gran solemnidad* [CARLOS is grabbed between the three and tied to a chair. HELENA is also tied to a chair and they put a handkerchief in her mouth. The atmoshphere is no longer violent, it seems to turn into an act of total submission. Everything is prepared with great solemnity]" (Act 2: 37). The care and solemnity with which the intruders proceed suggest that their actions form part of a ritual in which they are all participating and lend a more ominous feel to the entire episode. Similarly, the parallel between the stage and the outside reality becomes particularly uncanny at this moment, given the prominence that kidnappings and disappearances had and would continue to have throughout the 1970s in Argentina. Here, Pavlovsky is alluding to the political and social volatility offstage, making all the more poignant and violent the scene that has been constructed in the bourgeois apartment. The scene set, the four intruders begin to pull out a confession, any confession from Carlos in an effort to humiliate him:

> *Carlos*: ¡Les digo que no tengo nada que confesar! ¡No voy a confesar nada, porque no tengo nada que confesar!
> *Sueco*: (*Tocándole la cabeza muy cariñosamente*). Siempre hay algo que confesar siempre.

Carlos: ¿Qué quiere que le diga?
Sueco: Interrogalo. Turco.
Turco: ¡Bueno...! Yo voy a prender mi encendedor, y usted va a ser buenito, ¿no es cierto? (*Le pone el encendedor prendido cerca de la barbilla. ANÍBAL lo tiene agarrado de la cabeza. CARLOS grita de dolor*).
Sueco: ¡Vamos, Carlos! Usted es un buen muchacho y tiene que ser obediente. (*El TURCO le mete el pulgar en el ojo. ANÍBAL lo tiene agarrado. El SUECO le pega un rodillazo en la barriga. CARLOS está a punto de desmayarse*).
[*Carlos*: I tell you I have nothing to confess! I'm not going to confess anything, because I have nothing to confess!
Sueco: (*Touching his head affectionately*). There's always something to confess.
Carlos: What do you want me to say?
Sueco: Interrogate him. Turco.
Turco: ¡OK...! I'm going to light my lighter, and you're going to be good, isn't that right? (*He puts his lit lighter near his chin. ANÍBAL has him by the head. CARLOS screams with pain*).
Sueco: Come on, Carlos! You're a good boy and you have to be obedient. (*TURCO sticks his thumb in his eye. ANÍBAL is holding him. SUECO knees him in the belly. CARLOS is about to faint*)] (Act 2: 38).

In this scene, the tables turn from when Carlos appeared to control everything and the intruders are once again in command of the situation with Sueco directing. His comments and actions toward Carlos ("(*Tocándole la cabeza muy cariñosamente*). Siempre hay algo que confesar siempre [(*Touching his head affectionately*). There's always something to confess]") seem to indicate that he knows more about Carlos than Carlos himself knows, further emphasizing his own position of power over the others in the room. The participation of the three men in Carlos' torture underlines the quotidian-ness of physical violence (the tools they employ are simple and everyday) simultaneously with its extreme-ness (Carlos finds himself about to pass out from the pain). This contradiction within the intruders' violent acts becomes particularly poignant when Carlos finally confesses to the minor indiscretion of having masturbated at work:

Carlos: (*Desesperado cuando el SUECO se acerca con la pinza*). ¡Ayer me masturbé!
Sueco: ¿Dónde?
Carlos: En la oficina.
Sueco: ¿A qué hora?
Carlos: A las cinco. Estaba solo. A veces lo hago cuando estoy nervioso. Me metí en el baño, me masturbé y me tranquilicé. A veces lo hago para tranquilizarme.
Sueco: ¿Es cierto esto, Carlos?

> *Carlos*: Sí, es cierto.
> *Sueco*: Bueno, discúlpeme. Turco, curale el ojo y desatalo. Aníbal, pará el grabador. Flaco; desatá a la señora Helena. (*CARLOS tiene un ojo tumefacto. El TURCO le cura la herida y le pone una curita. Hace el trabajo con gran dedicación, durante la escena siguiente*).
> [*Carlos*: (*Desperate when SUECO approaches him with the pincer*). Yesterday I masturbated!
> *Sueco*: Where?
> *Carlos*: In my office.
> *Sueco*: What time?
> *Carlos*: At five. I was alone. Sometimes I do that when I'm nervous. I went into the bathroom, I masturbated and I calmed down. Sometimes I do it to calm down.
> *Sueco*: Is this true, Carlos?
> *Carlos*: Yes, it's true.
> *Sueco*: OK, excuse me. Turco, fix his eye and untie him. Aníbal, stop the recorder. Flaco; untie Miss Helena. (*CARLOS's eye is swollen. TURCO cures his wound and puts on a band-aid. He does the work with great dedication, during the following scene*)] (Act 2: 39).

Through Carlos' confession, the spectator comes to realize that the ritual that the four intruders enact upon the couple is a farce, one that is used to uncover the peccadilloes that everyone hides but also a manner to humiliate and shame the victims. Thus, the violence here is used as an instrument to inspire a certain reaction. Here, violence *inspires* fear in Carlos for the purpose of humiliating and breaking him and, then, subjecting him to the power and control of the others.

The purpose of violent gestures in *La mueca* is to gain control and power over the others, a position that the characters attempt to acquire throughout the play. This way in which violence is the means to a goal and the goal itself is interesting in light of Flaco's comments at the beginning of the second act. As we have seen, Flaco is the one who is set apart from the others, using intellect and creativity to rule the others rather than force. Sueco, seeing Flaco reading, asks what he is doing:

> *Flaco*: Pintura del Renacimiento. No te cansás de mirarlo...Arte sin violencia.
> *Sueco*: Otra época, Flaco, otros mundos, otros hombres. Yo preferiría no ser violento, pero estamos hecho añicos y tenemos que reconstruirnos, pedazo a pedazo.
> [*Flaco*: Painting from the Renaissance. You don't get tired of looking at it...Art without violence.
> *Sueco*: Another time, Flaco, other worlds, other men. I would prefer not to be violent, but we're shattered and we have to reconstruct ourselves, piece by piece] (Act 2: 26).

Both statements in this scene reflect the character of the man who utters them. Flaco's description reveals a desire to move beyond the quotidian violence in which he finds himself, whether that be by engaging another time or creating his own narrative. Sueco, in turn, embraces the reality in which he finds himself and attempts to engage it on his own terms, manipulating and recreating. These comments also allude to the contemporary moment in which *La mueca* is being written and performed, a time when violence defined human and political relationships and did not seem to abate. Both Flaco and Sueco, however, recognize the place that violent acts have in their time and existence, thus, saying that it is impossible to define one's self without violence.

One of the most important moments of *La mueca* comes in the first act soon after the couple enters the apartment when they are referred to as "Él" and "Ella." As seems typical for the play, the couple enters the room arguing over small details of the night, a practice which turns out to be a manner of foreplay as the man expresses in the following quote: "Qué linda te ponés cuando estás libidinosa. ¡Vení, juguemos un poco que ahora somos cuatro...! [You're so pretty when you're lustful. Come on, let's play because now there's four of us...!]" (Act 1: 21). The woman rejects his advance given that it is not the chosen night, only giving in when he promises to pay the decided upon rate. After handing her a check, the two begin their ritual:

> ELLA se suelta el pelo. EL se abre la camisa. Se esconde detrás del sillón. Aparece por un extremo en cuatro patas gruñendo como un león. ELLA lo mira un poco indiferente. Luego EL sigue gruñendo, ELLA se pone en cuatro patas y EL la sigue como si fueran dos animalitos. Juegan que EL la corre a ELLA. El juego es muy excitante. Juegan en cuatro patas remedando un león y una leona, un poco coqueteándose. Debe ser un juego habitual, mímico y bien hecho. EL la besa en el cuello, la huele, ELLA se deja oler el cuerpo. Cuando EL la quiere montar ELLA no se deja. Corren. La escena es estéticamente agradable. El juego es en serio. ANÍBAL por detrás emerge la cabeza y hace señas al SUECO como diciéndole: "sensacional." El SUECO le hace gestos afirmativos. El TURCO también saca la cabeza y se abanica como diciendo que la escena es demasiada erótica para aguantarla. La pareja sigue jugando. Desaparecen detrás del sillón. Los cuatro personajes miran atónitos desde sus escondites lo que debe ser el acto sexual, lo que el público no llega a ver. Y evidentemente no se cuidan tanto porque se presume que marido y mujer están muy concentrados en el acto.
>
> [SHE lets down her hair. HE opens his shirt. He hides behind a chair. He appears on the other side on four legs growling like a lion. SHE looks at him indifferently. Then HE continues growling, SHE goes down on four legs and HE follows her as if they were two animals. They play that HE is chasing HER. The game is very exciting. They play on four legs mimicking

a lion and lioness, flirting. It must be a habitual game, well done. HE kisses her neck, smells her, SHE lets him smell her body. When HE wants to mount her, SHE doesn't let him. They run. The scene is aestetically pleasing. The game is serious. ANÍBAL's head emerges in the back and he makes signs to SUECO as if saying: "sensational." SUECO makes affirmative gestures. TURCO's head also comes out and he fans himself as if saying that the scene is too erotic to stand it. The couple keeps playing. They disappear behind a chair. The four characters watch astounded from their hiding places what must be the sexual act, what the audience doesn't see. And evidently they don't take too much care because they assume husband and wife are concentrating on the act] (Act 1: 21–22).

It is important to note that the couple does not have the autonomy that will later come when they are named. Their individual selves are only defined by class differences. Here they are both bound by the generic "he" and "she" and their actions are dictated by a ritual in which they engage on certain nights—lacking any momentary innovation. Furthermore, the ritual is not even original to them. In this way, the couple is denied any freedom of thought or action and is identified as one of many, their actions stereotypical of a group. This formulaic presentation of the heterosexual sexual act is consistent with the presentation of characters that was examined earlier, but, at the same time, carries this one step further in equating the couple and their actions with animals—a pairing that is reiterated by the text when it refers to the couple as "animalitos."

Remembering this beginning it is interesting to note that the play ends with the intruders packing their bags and leaving. In their wake, the couple and the audience hear an earlier recording of Carlos and Helena. This ending leaves the couple broken apart by their confessed infidelities and mutual loss of respect of one another. The violence at the end, then, is not the physical beatings or insults that characterize the body of the play, but the shattering of identities that is left in the wake of the intruders' filmmaking.

Eduardo Pavlovsky's *La mueca* looks to challenge constructions of identity by revealing the performativity of social roles. The characters question definitions of class and gender through their violent actions and words. Violence here is used to create an identity that will exert dominance over others. Pavlovsky is showing how the performative roles that we choose to take on are created through a violence against others. This play is not written as a type of theater that aims to entertain and maintain the status quo, but instead within a tradition that questions its surroundings and provokes its audiences. *La mueca* wants to reflect the political and social context in which it was written. This means that the central aim of the play is to highlight violence and the way violence creates who we are. Violence here becomes a tool in which the characters create themselves and others in relation to this violence. Violent acts and the struggle for control create a world in which the characters can understand one another and themselves. This does not mean that this is a didactic theater that aims to teach its audience how to act. Instead, Pavlovsky provokes the audience's reflection on violence and its place in their everyday lives, and urges us to be aware of how our violent acts exert and maintain a dominance that perpetuates the cycle of oppression mirrored in political and national contexts. One of the elements that Pavlovsky underscores is the focus on the metatheatrical violence in terms of class, a move that will be intensified by Gambaro in terms of theatrical structures in *Información para extranjeros*, examined in the next chapter.

Chapter 4

Disorderly Conduct: The Violence of Spectatorship in Griselda Gambaro's *Información para extranjeros* (1973)

*Cuando nadie puede abrir la boca,
¿por qué se va a gritar gratuitamente?*

Griselda Gambaro, Información para extranjeros

Información para extranjeros (1973) by Griselda Gambaro is a play that fragments the theater experience into small pieces that portray violent acts taken from Argentine and international newspaper accounts. Gambaro's play is an avant-garde example of a theater that violently and unconventionally wants to show its spectators (and readers, in the case of this play) the worldview that it has discovered. Gambaro creates this alternative theatrical experience by stating that the play be performed in a space that is not a theater. The spectators move through the different rooms of a house, each accommodating a different violent episode. In this way, the role of the audience is changed from one that is physically detached to one that is intermingled with the action and the actors. Each group of spectators receives a guide who, like the Mendigo in Estorino's *La dolorosa historia*, leads them through rooms to watch violent kidnappings and episodes. The audience becomes what Milanés was: spectators traveling from scene to scene, unwilling witnesses forced to see violent images and memories. The foreigners that are alluded to in the title are all of the spectators (and readers) that come to *Información*, thus stating that no one is familiar with the material in the play. Spectatorship, for Gambaro in *Información*, becomes an experience of estrangement, of "foreign-ness." As foreigners, the spectators are drawn into the action unfolding before them but always with a level of detachment that allows them to question these violent scenes. This distancing, borrowing a term from Bertolt Brecht, provokes thought and it allows the spectators to intrude upon a violent, intimate moment and create a parallel between the onstage world and the violent, offstage one in which the play was written and to which it refers.

Griselda Gambaro's *Información para extranjeros* brings together performance and theater in a blurring of lines that demands an understanding of performance. While theater is undoubtedly comprised of both the text (the written script) as well as the representation of that script, what has come to be known as performance art should not be confused with theater. Gambaro's play has a written text that guides and dictates the actor's movements and words but also changes the traditional role of spectator to one that, through his/her very presence, innovates the spectacle with the possibility of intervention. The precariousness of the plot that the spectator's possible intervention lends to the play places it in a space where Gambaro can use elements from both genres to bring about the desired effect of the overall work. Despite the difficulty in defining performance, the instability of the representation rather than the adherence to the text distinguish Gambaro's *Información* from other theatrical plays and hint at the beginnings of performance art. Griselda Gambaro's *Información para extranjeros* engages with these different genres in order to change what theater is and to, thus, engage her spectators in both the spectacle of *Información* and the way the play is read. In this way, the violence that is the central subject of Gambaro's play mixes with traditional and innovative notions of spectacle and spectatorship to push the boundaries of her play and theater and the social context to which it is contributing.

Información para extranjeros is the most different in structure of the plays examined here. It radicalizes the role of the spectator and the actor within the theater by placing a strong emphasis on the former. The spectator will be as much in the figurative and literal spotlight as the actor. This change creates a text and representation that will change from place to place, performance to performance, and even from spectator to spectator. This is due, without a doubt, to the context in which Gambaro writes and her desire to intervene in the contemporary situation. As Susana Tarantuviez points out, this is in part a play denouncing the repression of the Argentine situation.[1] However, I also see a difference in perspective on theater that can be traced to the fact that the playwright is one of few women from a modest, immigrant past entering into a profession historically dominated by men and that demands a large economic investment to succeed. While this is not to say that Gambaro's personal history determines her work, it would be remiss not to reflect upon how an author's personal and historical circumstances influence her theater.

Griselda Gambaro can be considered one of the most important dramatists in Latin America in the twentieth century and without a doubt the most important woman dramatist in a field that has traditionally been difficult to break into for women.[2] Her list of plays from the 1960s and beyond, for example, *El campo* (1967) and *Los siameses* (1967), is extensive and central to the history of Latin American theatrical production.

Gambaro's theater is often focused on the relationship between the dominated and the dominator, an idea from which she somewhat departs in *Información para extranjeros*. Gambaro explores the role of the victim, concentrating in her early theater on the lack of will exhibited by the victim, often a woman. Moving into her plays from the 1980s and beyond, this role changed to become one where the woman discovers her own will and strength. This can be seen in *La malasangre* (1981), where Dolores, through the course of the play, comes to understand and challenge the perversity of the context around her. Connected to this are the themes of problems of communication and power games, topics that can also be seen in *Información*. The construction of memory and the connection to the reality of the Argentine situation are also constants in Gambaro's theater. An example of this is *Del sol naciente* (1982) where the question of land and who owns it refers to the context of *La Guerra de las Islas Malvinas* (Falkland Islands War). The allusion to the Argentine reality is perhaps the strongest theme that can be seen in *Información*, a play that serves as a testimony and a warning of what was happening in Argentina in the late 1960s and the 1970s.

Información para extranjeros can be seen to be, in part, a prophetic testimony of the contemporary events that were and would take place in Argentine politics throughout the 1970s. The play was written during roughly the same historical period of Pavlovsky's *La mueca*, a context that should be stressed and understood in reference to this play. The relative calm in the wake of the military coup d'état of General Juan Carlos Onganía in 1966 was breached in May of 1969 with the *cordobazo*. This event consisted of riots of students and automobile workers that broke out across the city of Córdoba and led to divisions within the army on how to regain control. Onganía saw this as the result of the influence of the Cuban revolution and wanted to use force against the rioters and the battling factions. General Alejandro Lanusse (who would later assume the presidency), on the other hand, wanted concessions, the idea that ultimately won out.

Meanwhile, the Argentine economy continued to deteriorate, contributing further to national and widespread dissatisfaction. As the government's grip unraveled, all of these factors converged to increase the number and power of the guerrilla groups in the country. Armed struggle began and the formation of leftist guerilla groups was quickly followed by others on the right. Kidnappings became more and more common as they were used as a political tool. The kidnapping and subsequent assassination of the former president General Pedro Eugenio Aramburu in March of 1970 was perhaps the most famous and the act that brought about the fall of Onganía's government.

General Rodolfo Levingston became president and issued the "Hora del Pueblo," a joint manifesto calling for the resumption of civilian rule. However, his tenure only lasted through the first few months of 1971, given the disruption in February of the *viborazo*, a second and seemingly more violent outbreak of riots in Córdoba. Lanusse, who had been operating in the background during the preceding years, assumed leadership duties. In Mendoza, riots broke out as protest to the increase in guerilla warfare and Lanusse attempted to shore up support among the unions. For a time, things went well, but then inflation rose again and Lanusse responded by lifting an eighteen-year ban on Peronism.

This action put into motion Juan Domingo Perón's eventual return to power in 1973, after the failed presidency of Héctor Cámpora. Cámpora's inability to govern was highlighted when fights erupted among Perón's supporters at Ezeiza airport in June of 1973 while waiting for Perón to arrive. However, when Perón did assume power at the end of the year, with his third wife, María Estela Martínez de Perón "Isabelita," as his running mate, he was unable to control the warring factions that found home under the wide umbrella of Peronism. The violence between the right and the left of the earlier years continued. On the left, the Montoneros and the Ejército Revolucionario del Pueblo (ERP) carried out numerous assassinations, kidnappings and robberies. On the left, a new secret organization called the Alianza Argentina Anticomunista (Triple A), said to be connected to the federal police, emerged on the political and social scene, instilling terror.

Initially, upon his return in 1973, Perón enjoyed some political calm and economic success. He tried to break with the leftist groups that he had earlier used to regain control. Stiffer penalties were issued against terrorists (but the Triple A was ignored) and he broke with the Montoneros and other groups in May of 1974. He died two months later from a heart attack and his widow assumed the presidency. With this, the precarious hold on order was irrevocably lost and violence and inflation returned with a vengeance. The Montoneros resumed clandestine operations after being rejected by Isabelita Perón. Guerrilla warfare increased and the Army was given virtually *carte blanche* to fight against it. More and more economic problems arose and, finally, impeachment proceedings were filed against Isabelita with the accusation that she had stolen money. On March 24, 1976, the president was abducted and the military moved forward with a coup d'état against the government of Isabelita Perón. With this action the *Proceso de Reorganización Nacional* [Process of National Reorganización] began, an action that is also known as the Dirty War. *Información para extranjeros* was written in 1973 (though it was not published until later), and contains episodes that testify to many of the violent events of the period. It is a play that connects strongly with its historical and political context in an effort to act as witness and to influence its reading and viewing public.

Información para extranjeros leaves behind the traditional notions of spectatorship in theater as well as that of a clear, linear argument that advances through the beginning to the middle and the end.[3] Instead, Gambaro divides the story into twenty scenes of unequal length that create an experience where the spectators are guided through rooms in which atrocities and disappearances are taking place. These episodes of violence that are being portrayed have been adapted from other texts—many being local and foreign newspaper articles. The purpose of this play is to force the spectator to see what is happening outside the theater. In that way, the violence on stage is directly linked to that which is taking place offstage. *Información* is a fundamental work that clearly establishes a link between the onstage text and the offstage reality. Gambaro wrote this play hoping to bring to light both in Argentina and outside the national borders the horrific social reality that she and others were experiencing in Argentina and that would continue for the rest of the decade. Nevertheless, it is necessary to point out that the play has not premiered in Buenos Aires, only in Mexico and Germany with the performance of some select scenes in the Argentine provinces. This, in many ways, indicates the connection between Gambaro's play and the immediate context in which it is written, suggesting that the play loses its validity when taken out of its temporal and historical context—an assertion that Gambaro herself put forth when asked about this play at the GETEA (*Grupo de estudios del teatro argentino* [Group of Studies of Argentine Theater]) conference in Buenos Aires in August of 2005.

It is difficult to define the argument of *Información para extranjeros* in a clear, direct way, such as that which was used to outline the narrative of the other three plays explored here. Gambaro's play is not one with a story that advances from beginning to end but is divided into twenty different scenes that portray episodes of violence which the audience will experience, an experience that will vary depending upon the day, the people, and the order in which they are seen. In this way, the idea of plot is violently questioned and transformed to fit the present situation.

In the very beginning, the spectators are divided into groups; each group receives a guide who will lead them through the twenty scenes in differing orders from the other groups with the exception of the final scene. This is the only one that Gambaro instructs all the groups will experience together and, in this way, makes it stand apart. Therefore, the written script that readers have (the one that is analyzed in this chapter) is just one of the many possible manifestations of what the spectator could experience at a production.[4] These scenes come from a variety of contexts (Argentine, German, literary, newspaper accounts) and contribute to the creation of a wide-reaching play that aims to interrogate its own historical and temporal context and can offer a window into the exploration of spectatorship in Latin American theater. *Información para extranjeros* is portrayed as an experience where each spectator's encounter will be individual, though with the objective that they will come together against the practice of violence and torture.

The play constructs itself as a tour of a sort of museum or fun house where the spectator participates in the exhibitions through a physical journey that makes stops along the way. Though the central argument of the play is hard to define given that the scenes should be experienced in different orders depending upon the route that the guide chooses for the spectators, the play writes itself against the violence and torture that it attempts to reproduce onstage by forcing the spectator-reader to confront the horrific events and the emotions that they will arouse. In this way, Griselda Gambaro creates in *Información para extranjeros* a dependency on space within the parameters of the play. The initial setting of the scene—a large, residential house—creates the atmosphere in which Gambaro unfolds the action of her work. The plot cedes its traditional position of centrality to that of space and, particularly, spectatorship. Both space and spectatorship, and the interaction of the two within the play, become the fundamental ingredients of *Información para extranjeros*.

It is important to remember that the "storyline" of the play's script is divided into twenty scenes; what's more, the full title of the play states it is a "Crónica de 20 escenas [A Chronicle of Twenty Scenes]."[5] This description of *Información* is particularly important when we look at the word's definition. The Dictionary of the Real Academia Española defines *crónica* first as "Historia en que se observa el orden de los tiempos [Story in which a timeline is observed]," and then as "Artículo perodístico o información radiofónica o televisiva sobre temas de actualidad [Article or information found in the newspaper or on the radio or television about current events]."[6] This second definition is particularly apt for the play in that it consists of twenty scenes that reproduce episodes of physical or psychological torture some of which had been reported in newspapers or other sources from the time. In this way, Gambaro is deliberately linking her work with contemporary acts and, in effect, attempting to reproduce a journalistic simultaneity.

Información para extranjeros is thus engaging with the audience about the contemporary moment in which it finds itself. The emphasis in the RAE on the medium that is used to convey the information (newspaper, radio, television) allows Gambaro's play to step outside theater and dialogue on issues that can be seen as *the* domain of other genres. *Información*, then, attempts a blurring of borders between genres that advocates an openness and an opening—of minds, of definitions, of lives, of political thought.

Similarly, the first definition found in the RAE dictionary lends itself to Gambaro's play in that she is recording for an audience a particular moment in the Argentine context that consists of various parts that must be brought together in order for them to be understood. At the same time, *Información* undoes this definition by allowing a multiplicity of valid orders of the scenes of the play. With this questioning, the play advocates a similar task in its spectators and readers—a questioning that will not allow the play to be represented or read without reflection on itself, on its context, or on the violence that it reproduces.

Gambaro's approach to her title is not the only place of innovation in *Información* that alerts the spectator to the play's originality. Inherent in the first definition above is the idea of the presence of an audience, a detail that proves essential to this play. The opening pages that set the scene also prepare the audience for a different theater experience that will test the limits of its spectators. First, there is no list of characters that sets forth the *dramatis personae*. Instead, there are detailed notes that attempt to outline the play's direction. The absence of a list of characters suggests that the play is defined not in a traditional way in which the characters carry the story, but through the action, and can be adapted according to the project to which each individual theater group attempts to allude. Similarly, the stage directions echo the flexibility found in the lack of character definitions. Gambaro's central project does not develop around a group of characters but instead around a group of experiences that the actors should try to bring about through the scenes. Space, then, takes on an importance within the play that can replace that of the characters. The space that Gambaro designates for *Información para extranjeros* is particularly interesting in that she moves her play out of a "traditional" theatrical space in order to place it in that of a large house that has an ample amount of rooms and hallways in which the action of the play can unfold. Gambaro designates the specific manner in which these spaces should be used and the objects to be found within them:

> *El ambiente teatral puede ser una casa amplia, preferentemente de dos pisos con corredores y habitaciones vacías, algunas de las cuales se comunican entre sí. Un espacio más amplio para la escena final.*

> *Ubicados en los pasillos, dos o tres cajones verticales, rectangulares, apoyados contra la pared, con puerta y respiradero.*
> *En un espacio a elección, un cajón un poco más grande de iguales características.*
> *Los corredores están a oscuras algunos y otros crudamente iluminados, en evidente contraste.*
>
> [The theater space can be a spacious, residential house, preferably two stories, with corridors and empty rooms, some of which interconnect. A larger space is needed for the final scene.
> Situated in the passageways, propped against the walls, are two or three vertical rectangular boxes, each with a door and air holes.
> In a different area, chosen by the director, sits an additional box, larger but otherwise the same as those in the passageways.
> Some of the corridors are dark, while others, in obvious contrast are crudely lit] (69; 69).

First, the play will unfold in a house, and not in a traditional theater in which the audience would be seated and separated from the actors who are on the stage. Thus, the indication from the beginning is that this play is not a conventional work but one that strives to create an experience in which the spectator will participate. Furthermore, placing the action in a house questions the play's theatricality and locates it in a much more real space that directly connects with the spectators' everyday lives. Gambaro indicates that in the hallways that will connect the various rooms there should be certain props—large boxes with doors, like a sort of locker. Their importance is stressed given the detail that is used to specify their location and characteristics. In this way, they become an essential point of discussion—even more so later on when, in one of the scenes, a man is discovered in one of these lockers. Gambaro creates from a simple *cajón* another space, thus inspiring action and, from there, discussion in a way that characters do in Estorino's *La dolorosa historia*. In this way, space as well as the manipulation of space takes on a central role.

Space is an important point of discussion in *Información para extranjeros* given the avant-garde role that Gambaro assigns to it. It becomes a central aspect that contributes to the play's purposes and objectives. As seen in the quote above, Gambaro takes a more quotidian location and converts it into a theater. With this, the play blends the non-theatrical world with the theatrical and begins to connect them in a way that will question which is which and where the spectator lies. Though the manipulation of the theatrical space is a central issue to much modern drama, Gambaro's play takes this process one step further in her innovation of the role of spectator and space. Here, Gambaro uses it to create a feeling of confinement and separation in the spectator that multiplies the effects of the central plot. As

Rosalea Postma has stated, the involvement of the spectator in the central action through the absence of seating as well as the manipulation of large spaces into smaller, more claustrophobic ones, contribute to the creation of a space in which the spectator cannot detach him/herself from the action and is inescapably drawn into the play.[7] In this way, the management of space in the play is used to increase the desperation that the episodes portray and that *Información* pushes even further through its innovation here. This is essential to discuss in the play given that Gambaro's intention is to take it out of a traditional theatrical space and place it somewhere else and, thus, use theater to engage with a larger audience of both spectators and readers.

This avant-garde use of space is an element that reminds us of Antonin Artaud's thoughts on theater. In Susana Tarantuviez's view, it is this innovation of space in *Información* that is precisely one of the ways Gambaro adheres to Artuad's ideas.[8] Tarantuviez recalls how Artaud advocated a change in the traditional set up of the theatrical space, such as the actors performing around the audience. This shift to where the spectator occupied the space in the middle of the action would create a situation where the focus was not exclusively on the actors since they would not be the center.[9] Gambaro takes advantage of Artaud's theories by using the rooms in a house and by mixing actors and spectators. Like Artaud envisioned in his Theater of Cruelty, Gambaro wants to question these borders between the theater and reality, between actors and spectators.

Patrice Pavis in his *Analyzing Performance: Theater, Dance, and Film* details the importance of space in accordance with two different theories: empty space to be filled or unlimited space that is to be expanded. This distinction is interesting in reference to Gambaro's *Información* given the centrality of space in the play. Whereas in Estorino's *La dolorosa historia* the play's space was to be filled, an essential act that had reference within the argument of the play, Gambaro's space is much more wide-ranging and expansive. It has no limits within the plot of the play but instead seems to spread out into the context it has. The Cuban play fits into Pavis's first theory of a space to be filled, and the Argentine example is the second, given that there is such a flexibility that is central to the play's argument. This flexibility is essential to Pavis's definition of space and emphasizes the meaning that the actors give. This increases in the case of *Información* to include also the spectators since their role becomes central to the plot. Space, then, in *Información* becomes limitless; its meaning expands to encompass all of the events that are happening in the episodes but also those that could happen. It is not defined or contained and it is this aspect that permits Gambaro's play to have such a chilling effect on her audience, an effect that further affects the message of the play.

The role of spectatorship similarly occupies central stage in *Información para extranjeros* given that Gambaro specifies the construction of the audience that will view the play. As can be seen above with the management of space in the play, those spectators become an integral part of the development of action—not detached physically from the stage, but intermingling with the actors. In the stage directions quoted below, Gambaro details how spectators are divided into various groups and how the argument, or better yet arguments, of the play will develop, sometimes independently and sometimes concurrently, for and with these individual different groups:

> *El público será dividido en grupos, cuyo número y cantidad dependerá del espacio con que se cuenta. La identidad de cada grupo puede facilitarse con un distintivo numerado o de color.*
>
> *El grupo 1 marca uno de los posibles desarrollos de la acción. Los guías 1, 2, 3, 4, etc., conducen sus respectivos grupos. El desarrollo de la acción que observan estos grupos es aleatorio hasta la Escena final Nro. 20, donde todos los grupos convergen.*
>
> *En determinadas escenas, los actores actúan como público, del cual no se distinguen en principio, pero el público no será nunca forzado a participar en la acción.*
>
> *Los grupos se entrecruzan en los pasillos, pueden contemplar la misma escena, ya sea en los pasillos o cuando el director lo estime necesario.*
>
> *Los textos enunciados por los guías como "Explicación: para extranjeros" han sido extraídos de los diarios de la época.*
>
> [The audience will be divided into groups, the number and size of which will depend on the space. A particular number or color can serve to identify each group.
>
> Group 1 will mark one possible development of action.
>
> Guides 1, 2, 3, 4, etc., lead their respective groups. The order in which the scenes are observed by these groups is left to the director's discretion until the last scene, scene 20, when all groups converge.
>
> In certain scenes, actors play audience members and are actually part of the audience. Audience members, however, are never forced to participate in the action.
>
> The groups cross in the passageways—when the director considers it necessary.
>
> Excerpts introduced by the guides as "Explanation: For Foreigners" come from Argentine newspapers of the period 1971–72] (69–70; 69–70).

Gambaro indicates that the audience will be divided into separate groups and, that the development of the play will unfold along similar but different lines. *Información para extranjeros* attempts to mirror the process of life lived offstage, where the spectators view different realities and in varying order. The experience attempts to reproduce a similar reality through diverse paths, demonstrating the differing realities that exist both on and off stage. At the same time, Gambaro specifies that the last scene will be viewed by all the groups simultaneously, underlining the play's desire to uncover a hidden reality that can be seen in the play's immediate context. As Dick Gerdes points out, the script of the written text of *Información* is just one possibility of many.[10] This multiplicity of experience and action development points to the closeness between this play and the genre of performance. However, Gerdes, while recognizing the role of improvisation, underlines how the play differs from the common experience of a 1960s "happening" through its careful planning and script development.[11] Nevertheless, it is important to note the role that performance has in this random presentation of scenes. Performance is defined by its unfinished quality, by the use of collaboration in every individual representation. *Información* is a play that presents itself as fragmentary in order to engage both sides—the actors and the spectators—in this unfinished element and the audience's ability to intervene in the presented events.

The play opens with the guides dividing the audience into groups, each of which will be led by a separate guide. The performance begins with these guides creating their groups and outlining for the spectators the work that is about to be seen. Though it is not specified this way in the written text, this part can be understood as the first scene of the play given that the guide's verbal introduction underlines the unorthodoxy of which the audience will partake and sets the tone for what sort of work this will be:

> *Guía*: Señoras y señores: la entrada es de mil pesos, para adultos. Si ya pagaron, nadie puede arrepentirse. El gasto ya está hecho. Mejor gozar. El espectáculo es prohibido para menores de 18 años. También prohibido para menores de 35 y mayores de 36. El resto puede asistir sin problemas. Ausencia de obscenidad y palabras fuertes. La pieza responde a nuestro estilo de vida: argentino, occidental y cristiano. Estamos en 1971. Les ruego que no se separen y permanezcan en silencio. Atención con los escalones.
>
> [*Guide*: Ladies and gentleman: Admission is ____, for adults. If you've already paid, you can't repent. The cost is already incurred. Better to enjoy yourself. No one under eighteen will be admitted. Or under thirty-five or over thirty-six. Everyone else can attend with no problem. No obscenity or strong words. The play speaks to our way of life: Argentine, Western, and Christian. We are in 1971. I ask that you stay together and remain silent. Careful on the stairs] (70; 71).

This preliminary introduction helps to set the scene for the play; what is most interesting is that the audience is the single most important thing that needs to be prepared and situated for the action that is to come. This preparation is done by indirectly outlining various aspects found in *Información*. The guide's introductory words underline the layers of spectators and the absurdity that the play is attempting to unveil. First, by prohibiting the play for anyone who doesn't fit into the age requirement, Gambaro makes implicit the fact that no one is suited for the material that is put forth. Nevertheless, maintaining the absence of obscenities or strong words questions the type of material that wouldn't be appropriate. With this assertion, Gambaro suggests that the idea of indecency or inappropriateness goes beyond a verbal vulgarity that the audience already knows. Here rather, as will be seen in the action following these words, there will be a physical inappropriateness and violence that will question our accepted notions of vulgarity and force the public to reconsider these definitions.

The connection between the material presented and the Argentine way of life in the above quote suggests that the definition of what is Argentine is linked to the violence that will be observed by the audience in the play. As Rosalea Postma has observed, by linking the violent material with Argentine self-definition, Gambaro's play questions how this society views itself.[12] She insinuates how acts of both physical and psychological violence have crept into the national definition. The verbal warning puts the spectators (and the readers) on guard to the unorthodox episodes and interpretations that will form the central presentation of the material.

Furthermore, the underlining of this being the Argentine way of life highlights the connection in contemporary Argentina between violence and the quotidian world. Gambaro is stating that the episodes of torture and violence that the audience will see in the upcoming scenes are part of the Argentine existence, embedded in the national definition. This reminds us of the change in the perception of war and politics that Idelber Avelar identifies in *The Letter of Violence*. Avelar traces the idea of war from the Chinese Sun Tzu in the fourth century BC through to Carl von Clausewitz of the German nineteenth century and beyond. Clausewitz maintained that all ethical consideration should necessarily end with war. As discussed by Avelar, Michel Foucault in the twentieth century, on the other hand, stated that politics was a continuation of war.[13] What is important to note for the Argentine society of the late 1960s and the 1970s is that this distinction between politics and war was being eroded. Violent war-like acts were being increasingly used as political responses between guerrilla groups and against Argentine citizens. Identifying the contents of the play as Argentine highlights this fusing between violent acts and political deed and forces the spectators to consider what this means for the definition of the country and its people.

However, the guide's introductory words to the spectators do not appropriately prepare them for the material they are about to see, and ultimately prove that the guide, either intentionally or inadvertently, is not trustworthy. His/her words do not agree with the material that is subsequently presented. As is observed throughout the play, the guide leads the spectators on erroneous paths. The first scene presented in the written script is a short one that reveals just how unreliable the guide is and how the spectators will need to make their own connections and suppositions about what is being presented and, most importantly, from what is not visible. In this scene, the guide leads his/her group through the hall to a room that is completely in the dark. The door shuts behind them and a muffled conversation is heard—all signs that things are not completely right. Nevertheless, the guide begins to speak as if the situation before them all were normal, simply mentioning to the spectators what can happen under the cover of darkness and never alluding to the fact that he/she has perhaps led them astray or the scene is somehow amiss:

> *Guía*: Un momento. No encuentro mi linterna. Recuerden que la ocasión hace al ladrón. ¡Cuidado con los bolsillos! (*Ilumina una pared rugosa y oscura*) Sólo han quedado las paredes desnudas. (*Corre la luz*). *Un hombre está sentado en una silla, como única prenda lleva un taparrabo de color desvaído. Alza la cabeza, sorprendido y atemorizado. Se cubre el sexo con las manos*)
> *Guía*: Perdonen. Me equivoqué de habitación.
>
> [*Guide*: One moment...I don't find my flashlight. Remember, opportunity makes the thief. Watch your pocketbooks! (Light comes up on a dark and wrinkled wall.) Only the naked walls are left. (The light travels. A man is seated on a chair, wearing only faded underwear. He raises his head, surprised and frightened. He covers his sex with his hands. To the audience.) Excuse me. I've got the wrong room] (1: 71; 71).

First, the darkness and the muffled conversation signal that something is happening secretively, something that cannot or should not be seen by the audience, and the guide, the one who is to escort them through this maze, leads the audience there despite its apparent inaccessibility. The simultaneous warning that the guide gives to the spectators to be on guard against pickpockets attempts to inspire fear in the audience for their own safety and distrust of a leader who would escort them into danger. At the moment, then, the spectators are distracted from the muffled voices and what they could be doing to instantly think of their own possessions and security. With the guide's warning, it can be imagined that the spectators mentally, and perhaps physically, cling more tightly to their purses and wallets. Following this invisible action, the guide begins to illuminate slowly the room in which they have entered. At first, it seems to be a plain, empty room. Then the light finds a semi-naked man who has been surprised by the group's entrance. His reaction shows his terror and fear, suggesting that the group is violating his privacy and space and hinting at past abuses and the fear that they will be repeated with this new group's arrival. Gambaro's use of this image for the first scene sets the tone of violation and terror that will accompany the entire play and similarly puts the audience on guard that the situation before them is one in which they will need to defend themselves without the expertise or help of a guide. The spectator is ultimately alone.

The guide ends this scene by ushering his/her group quickly out of the room, an act that insinuates that he/she has made a mistake by letting them enter here. By bringing to light the guide's unreliability in this scene, the spectators begin to question more deeply the guide's role in the play. The guide has proven him/herself to be unable to fulfill the job appropriately. He/she does not seem to know the layout of the house, where to enter, how to move forward. As a matter of fact, the second scene consists of the group moving to another door that is locked, inhabited by a voice that scorns the guide's desire to enter with his group: "*Voz (muy grosera)*: ¿Y a mí qué me importa? ¡Rajen! Estoy ensayando [VOICE: (very rudely) What's it to me? Beat it! I'm rehearsing]" (2: 71; 72). Given the evident incompetence of the guide, the spectators are left feeling abandoned in this play—a nightmare where each dark corner and locked room can hold a verbal or physical attack. They are unable to exit given the lack of light and orientation in the space and instead must rely upon this guide who consistently misleads them. In this way, the spectators are drawn into the representation given their close proximity to and their reliance upon the guide and the actors—they cannot detach themselves mentally from the situations before them the way a more traditional audience can because of their physical closeness to the action and their apparent need to guide themselves through the hidden dangers of the play. Rosalea Postma aptly identifies these feelings of helplessness and abandonment as the objective of Gambaro's play: "One of the functions of vanguard theatre according to Gambaro is to demythify the concept that the spectator has of himself. She executes her subversive thrust in *Información* as the spectator, unable to withdraw from the action, becomes a necessary part of the events."[14] The spectator becomes a central part of the play's storyline rather than a passive observer of the action as is seen in a more conventional theatrical experience. Although Gambaro states in the introductory stage directions that the audience will never be forced to participate,[15] her objective is to make the spectators change their definitions of their role within the play and, thus, change their experience in and of the performance. This renewal will, in turn, bring home the objective of the play: reconsider the role of violence in our lives and our own contribution to that violence. In this way, the physically present spectators watching a representation of *Información* are not the only intended audience. There are also the readers of the written script—a possibly large number given the fact that Gambaro's play was not published or produced in Argentina at this time.[16]

The spotlight on the spectator and the spectator's role within *Información para extranjeros* is particularly important for the play's argument and for the violent framework within it. The innovative role of the spectator found here connects in many ways with the innovative use of space. The spectator becomes implicated within the development of the play's action in a way that begins to blur the lines between the actors and the spectators. Gambaro's inclusion of actors within the groups of spectators as stated in the earlier quote underlines the link between theater and "reality" that this play is attempting to highlight and question. It is this connection that is being brought into the forefront of the discussion in order to force the spectator to recognize the actions offstage through the episodes of violence. Actors are among the audience members during the play's events and are thus incorporated into the play's direction and argument, as can be seen in Scene 6. In this scene, the guide and his/her group have just left one room and are changing floors, when the guide realizes he/she has led them astray and they all retrace their steps. The scene, then, takes place in the in-between space of the stairs, creating a liminal space that nevertheless becomes a vital element in the storyline. Suddenly interrupting the audience's movement from one room to another, a group of men grabs hold of one of the spectators to carry him/her away. What is most important to notice in the quote below is the lack of dialogue and the large amount of stage directions that indicate what is to happen in this scene. Despite the length of the quote, the scene would be acted out in a violently quick manner:

> (*Bajan. De pronto, bruscamente, irrumpe un grupo de hombres que se abalanza sobre una persona, mezclada con el público, que conversa con otro. A puñetazos y empellones lo llevan escaleras abajo. El otro permanece paralizado por la sorpresa un segundo, luego grita, precipitándose.*)
> Hombre: ¡Déjenlo! ¡Déjenlo! (*Consigue liberar al apresado. Los dos suben unas escaleras nuevamente, pero el grupo de hombres se abalanza hacia ellos, los rodean y los llevan arrastrando escaleras abajo*)
> (Se escucha por los altoparlantes una voz angustiada que dice.)
> Dios mío, ¿por qué corrí?
> (*Casi instantáneamente, con otros personajes, la escena se inicia y se repite en otro lugar. Los grupos pueden cruzarse en este momento. Se oye nuevamente la voz.*)
> Dios mío, ¿por qué corrí?
> (*La escena se repite en otro punto*)

Dios mío, ¿por qué corrí?
[(They go down. Suddenly, a group of men burst in, hurling themselves at a person in the audience who is talking with someone else. This other person is for a second paralyzed with astonishment. Then shouting, he throws himself into the fray.)
Man: Let him go! Let him go!
(He succeeds in freeing him. The two make it down a few stairs, but the group of men rush them, surround them, and drag them down the stairs. Over the loudspeaker a distressed voice is heard.)
Voice: My God, why did I run? (Almost instantaneously, the scene breaks out in another place with other characters. The groups may cross at this moment. Again the voice is heard.) My God, why did I run? (The scene is repeated in another spot.) My God, why did I run? (6: 87; 88).

This is perhaps the scene in which the role of the spectator-turned-actor becomes the most salient and crucial to the argument of the play. Here two 'spectators' are kidnapped in the middle of the play. This abduction is particularly violent and realistic in that the other spectators—the physical spectators of the live representation—are left feeling violated and vulnerable. In this way, Gambaro brings home the brutality and violence inherent in any kidnapping, whether it be of someone who is close in proximity or in sentiment. Similarly, the repetition found in this abduction—one happens quickly right after another—echoes both literally and figuratively its effects in that the spectator cannot break free from the atrocity but must re-live it *and* re-live the fear that accompanies it. The phrase that is repeated again and again over the loudspeaker—"Dios mío, ¿por qué corrí? [My God, why did I run?]"—demonstrates the futility of resistance that both the victim and the bystander begin to feel.

The written text mirrors the violent performance through the use of stage directions. These italicized directives interrupt the reader's movement through the action of the scene, truncating the reading experience as well as the action. In this way, the written text reflects the violent interruption of the performance and attempts to connect the experience onstage with the confusion and chaos offstage. This representation of a kidnapping in the 1973 text is both a remembrance of what was happening in Argentina during the tumultuous time after the coup of 1966 and what would come to pass within a few short years later when the *Proceso de Reorganización Nacional* came to govern in 1976 when disappearances of people in the streets was a common occurrence.

As seen in this particular scene, Gambaro abandons the more traditional role of the spectator as a passive receptor of the play's plot and purpose. The audience becomes an integral part of the story's development and its larger objectives outside the theatrical space. *Información para extranjeros* requires the spectator to enter into the play's direction given the various ways in which it implicates the audience within its narrative. In this way, Gambaro's use of violence reaches much further than being simply an observation of a spectacle. Instead, it becomes the spectacle into which the audience is drawn; their role is central to the violent dissemination in the play. Borrowing the idea of the *lector cómplice* [reader-accomplice] from Julio Cortázar, Jason Cortés identifies this as the complicity of the spectator within *Información para extranjeros*:

> En *Información*, Gambaro nos sorprende con la presencia de un espectador cómplice, pues no solo los personajes toman parte en la representación de la violencia, sino que los espectadores, a modo de perversión voyeurística, van a disfrutar del espectáculo. En este sentido, la obra nos somete a la violencia del conocimiento, donde el espectador asiste a un espectáculo del que no quiere ser parte, pero al que está obligado a participar
>
> [In *Información*, Gambaro surprises us with the presence of a spectator-accomplice, since it is not just the characters that take part in the production of violence, but also the spectators, in a voyeuristic perversion, who are going to enjoy the spectacle. In this way, the play subjects us to the violence of knowledge, where the spectator attends a spectacle in which he/she does not want to be a part but is obligated to participate].[17]

For Cortés, the spectator becomes an accomplice. This complicity contributes to the work's overall project to blur the boundaries between spaces. With this particular confusion of borders, Gambaro places the spectators in an awkward position in which they are meant to identify their own complicity with these violent acts. In this way, the central intention of the play takes place within the workings of the audience's physical and emotional reactions. Gambaro urges the spectators to question their own complicity with these violent acts and the other acts that reach beyond the theatrical walls of *Información para extranjeros*.

For the spectator-reader coming to *Información*, then, it is not enough to simply not commit violence against others; the audience is forced to see how inaction contributes to the creation of violence. The scenes that are portrayed in the play ask for an *espectador cómplice* [spectator-accomplice] who will make the connection between onstage and offstage disappearances, violent acts, and murders. From this connection, Gambaro hopes, the spectator-readers will act upon their horror and stop what is happening. Just as for Artaud, a spectator would be unable to recreate a violent act offstage after having seen it onstage, Gambaro's reader-spectators should be horrified by the violence they see and spurred to act against it.[18] In this way, we can see not only the amplification of the *extranjeros* in the title, but also foreigners in the more conventional meaning of the word—those living outside of Argentina who could more safely put pressure on the Argentine government to gain control of the violence that was defining the nation.

This emphasis on complicity between the spectator and the action represented seems to become most clear in scene four where Gambaro portrays the famous Milgram experiment that was conducted in Munich, Yale, and Princeton. Here, Gambaro makes a connection between obedience and complicity that provokes in the spectators and the readers a self-examination of our own role within the political and state violence unfolding before us. The findings of what has come to be known as the Milgram Experiment were published in 1963 in an article and in 1974 in book form. They took place originally at Yale as a way to measure people's obedience to authority and how far it would take them when they were asked to carry out corporal punishment on "victims" who answered wrongly. Stanley Milgram, the social scientist who carried out these experiments, discovered that obedience has in some ways been more deadly in human history than rebellion. Milgram set up a deceptively simple experiment: "A person comes to a psychological laboratory and is told to carry out a series of acts that come increasingly into conflict with conscience. The main question is how far the participant will comply with the experimenter's instructions before refusing to carry out the actions required of him."[19] In this experiment, two people come to the laboratory: one is designated as the teacher, one the student. They are told that the experiment is to test "the effects of punishment on learning."[20] The student is given information on which he will be tested.[21] He will receive an electric shock of increasing intensity when he makes an error. However, the student is really an actor and the focus of the study is the teacher, the one who is to administer the electric shock. In actuality, there is no electric shock being given: "The point of the experiment is to see how far a person will proceed in a concrete and measurable situation in which he is ordered to inflict increasing pain on a protesting victim. At what point will the subject refuse to obey the experimenter?"[22] The actor responds to the shocks accordingly and, thus, the experiment tests how far the teacher will obey the authority that tells him to press on rather than the manifestations of physical pain pronounced by the "victim."

Gambaro's purpose in including this experiment within the play is to highlight the way that ordinary human beings can become caught up in the complicity of violence. In the theatrical representation, the guide leads the spectators into a room that is connected to another where three men (Coordinador, Hombre maduro, and Joven [Coordinator, Mature Man, Young Man]) are awaiting their arrival in order to begin the spectacle. The rooms resemble a scientific laboratory complete with white mice in a cage and a white lab coat waiting for its occupant. The Coordinator of the experiment explains to the audience what they are about to see:

> Coordinador (al grupo, con tono profesional): Señores: nuestra experiencia tiene por objeto determinar el efecto pedagógico del castigo. ¿En qué medida el castigo acelera el proceso del aprendizaje? Imagínense. Si con una bofetada un chico aprende a andar derecho, nosotros dilapidamos años en enseñar y persuadir sólo con buenas palabras. No se puede perder tiempo. Ya será adulto, ya estará formado. Formado para la destrucción, cuando una bofetada, dos o tres descargas eléctricas en el momento justo hubieran puesto las cosas en su lugar. (*Comienza a observar al hombre maduro entretenido con los ratones*) Los señores nos ayudarán a esclarecer los...puntos...oscuros...¡Señor, déjese de jorobar con los ratones! ¡Idiota! (*Va hacia él y lo aparta a patadas*)
> Hombre maduro: Sí, sí. Perdóneme. Son tan lindos que...
> Coordinador (sosegado): Claro que son lindos. (*Se irrita*) ¿Y si empezamos?
> Hombre maduro: ¡A sus órdenes, señor!
> Coordinador (feliz): Patada va, asentimiento viene. Usted, señor, emocionalmente más maduro, será el maestro.
>
> [Coordinator: (to the group, in a professional tone) Gentleman: The subject of our experiment is to determine the pedagogical effect of punishment. To what degree does punishment accelerate the learning process? Imagine. If with one slap a child learns to behave, we waste years teaching and persuading only with nice words. We don't have time to lose. Soon he will be an adult; soon he will be molded. Molded for destruction, when one slap, two or three electrical jolts at the right moment could put things in place. (He begins observing the Mature Man playing with the rats.) The gentlemen will help us to clarify...unclear...details...Please, sir, stop pestering those rats! Idiot! (He goes toward him and kicks him away from the cage.)
> *Mature Man*: Okay, okay. I'm sorry. They're so cute that...
> *Coordinator*: (calm) Of course they're cute. (becoming irritated) Shall we begin?
> *Mature Man*: At your orders, sir!
> *Coordinator*: (happy) One kick...and acquiescence. You, sir, emotionally more mature, will be the teacher] (4: 74–75; 75).

The coordinator explains the objectives of the science experiment and the audience is given a sneak preview of what they will observe in his treatment of the Hombre maduro. When the older man begins to play with the mice and distract himself from the Coordinator's introduction, the Coordinator brings him mentally back to the task at home with a swift kick that is meant to reprimand and to show who is in charge. His attitude toward the Hombre maduro changes from that which he presented to the audience. The Coordinator becomes the leader, a status that is proven through physical force and through the Hombre maduro's recognition of him in that space. With this, the tone for the experiment is set and the roles are given out: the Hombre maduro will be the teacher and the Joven will be the student.

After receiving their money and signing release forms, they are separated into different rooms where the Hombre maduro puts on the lab coat and the Joven is stripped and tied to the chair. When he is warned that he can turn down the position, the Joven answers, "¡Si lo hago por la ciencia! [For the sake of science // Let us commence!]" (4: 76; 76). And with this statement, the experiment begins. The Joven is to answer certain questions that the Hombre maduro poses to him. When he answers incorrectly the Hombre maduro issues an electrical jolt that increases each time successively to 450 volts, the dosage that will cause death. The two men progress through the experiment, beginning with small dosages that increase quickly. As the experiment advances, the Joven begins to answer incorrectly and becomes more and more confused. Though the Hombre maduro asks if he should continue or if he should pause, he never questions the Coordinator's orders or authority over the situation, but instead remarks on the Joven's inability to keep the information straight. Here we see what Elaine Scarry has characterized as the doubting of someone else's pain, the impossibility of knowing another's pain. The Joven's response to the shocks are whimpers, cries, and sometimes screams. He cannot give words to his pain and, for this reason, the inflictor, here the Hombre maduro, cannot understand it. Just as Scarry states, there is doubt of the depths of another's pain: "Whatever pain achieves, it achieves in part through its unsharability, and it ensures this unsharability through its resistance to language."[23] The Hombre maduro cannot understand the pain that he is inflicting upon the Joven and, for this reason, continues with his actions.

The Hombre maduro persists in asking questions and issuing electrical shocks for the boy's mistakes until the 450 mark where the Joven is killed from the shock, even though he has given the correct answer: "*Maestro (sin mirar la lista)*: Se equivocó. Se equivocó... otra vez. (*Abre los ojos*) Es deliberado. Imposible que no sepa. Pero... me da pena... ¿no? (*Lentamente, aprieta el último botón de la caja. Silencio. Sonríe con alivio*) No gritó. [TEACHER: (without consulting the list) He made a mistake. He made a mistake... again. (He opens his eyes.) It's deliberate. He can't not know. Still... it hurts me... (He slowly pushes the last button on the box. Silence. He smiles with relief.) He didn't scream]" (4: 83; 84). With this final push of the button, the Hombre maduro ends the experiment and the Joven's life. The Coordinator, acting like a guide within the scene, accompanies him to the exit while he rationalizes that it is the Joven's fault for not knowing the answers. The Coordinator then turns to the audience to explain the experiment and its objective:

> *Coordinador* [...] *(Le estrecha de nuevo la mano. El Maestro sale. Coordinador, se vuelve hacia el público, profesional)* Esta experiencia, con los gritos grabados y las torturas simuladas, se repitió ciento ochenta veces. Desgraciadamente, este primer maestro que continuó los castigos hasta los cuatrocientos cincuenta voltios que determina la muerte, no constituyó una excepción. El ochenta y cinco por ciento de los maestros procedió en la misma forma. El mismo test se realizó en los Estados Unidos. ¿Los resultados?: sesenta y seis por ciento. Obedecían reglas y no eran responsables. Curioso ¿no? ¿Asombrados?
> *Guía*: Bueno, basta. No me aplaste el público. (*A su grupo*) La experiencia se realizó en Alemania y en Estados Unidos. Entre nosotros sería completamente absurda. Señoras y señores, busquemos algo más divertido. (*Conduce al grupo fuera de la habitación*) Por aquí, por aquí. Si son tan amables... Señoras y señores...
>
> [*Coordinator*: [...] (Again he shakes his hand. The TEACHER exits. The Coordinator turns toward the audience, professional.) This experiment, with recorded screams and simulated tortures, was repeated 180 times. Unfortunately, this teacher who continued his punishments to the lethal 450 volts was no exception. Eighty-five percent of the teachers proceeded in the same way. The same test was done in 1960 in the United States. The results? Sixty-six percent. They were obeying rules and weren't responsible. Curious, isn't it? Surprised?
> *Guide*: Okay, enough. Don't wear out the audience. (to his group) The experiment was done in Germany and the United States. Here among ourselves, it would be unthinkable, absurd. Ladies and gentlemen, let's look for something more amusing. (He leads his group out of the room.) This way, this way. If you would be so kind... Ladies and gentlemen...]
> (4: 83–84; 84–85).

The experiment is revealed to be not an exploration of the learning process and corporal punishment but instead a study to see how far people will follow orders. It is an experiment that is only possible through the Hombre maduro's lack of empathy for the victim, a circumstance to which the Coordinator attempts to draw the spectators' attention. The guide, however, quickly cuts him off and sets apart his audience from this experiment by saying that among the present context of people, the results would be completely different. He ushers his spectators to another room and experience, closing the dialogue on what he sees to be irrelevant. Gambaro's objectives, nevertheless, hide in these last words uttered by the guide. She aims to question how all of us believe that we would be different given the same circumstances, that *we* would not become torturers, thus obligating her circle of spectators and readers to question their own reactions. As Cortés observes, the appropriation of the experiment in this context suggests that any of us could become the Hombre maduro and simply follow the orders given us by the authority in the white lab coat.[24] Gambaro wants to push her spectators and readers to a space where we all examine our own actions and potential. In this way, this scene—scene four, early enough to set the tone for the reader—provokes in the audience a reaction where they will make connections between what is being represented here and what is being alluded to outside the scene. The *espectador cómplice* that Cortés recognized will see the older man's actions and how he quickly and unquestioningly follows the coordinator's orders and be horrified by the deafening silence that the final electric shock produces. And though the spectators and the readers may think he is a simple-minded man who respects all authority, the coordinator's closing words—that 85% in Germany and 66% in the U.S. chose the same path—will force them to think about what they would do and what their neighbor may choose to do. Moreover, the Guide's dismissal of the results becomes even more frightening given that the words show a refusal to understand the larger implications, thus setting the scene for much more serious results and implications. With these words, the readers and the spectators are forced to confront what they may be doing to contribute to the violence and, with this self-examination that the scene has provoked, Gambaro underscores the role of complicity in silence.

Continuing with the idea of complicity, the third scene of the *Información* continues to play on the connection between psychological and physical violence that is present throughout Gambaro's work. Here the action takes place in another darkened room and once again the guide introduces his group to the strange scene that is about to progress before them. It is important to note how the scene is set up by the guide before, and how he shifts to become a part of, not the event that is unfolding, but the audience that is partaking of the story:

> Guía *(gentil)*: Pueden ubicarse donde gusten. Las sillas alcanzan para todos. *(Mira)* No, no alcanzan. *(Las acomoda y ofrece)* ¡Preferencia para señoras...!
> *(Se ilumina la parte central del cuarto. Sentada en una silla hay una muchacha con las ropas empapadas. A su lado, de pie, un hombre la observa con una sonrisa de ternura. El Guía espera que la gente termine de acomodarse, señala lugares. Luego con un dedo sobre los labios, indica silencio, y se vuelve, como un espectador más, hacia los personajes que comienzan la acción).*

[*Guide*: [...] ([...] nicely) You can position yourselves wherever you like. There are chairs for everyone. (He looks.) No, not enough to go around. (arranges them, offers) Ladies first...!
(Lights on in the middle of the room. A young GIRL sits on a chair wearing clothes that are soaking wet. A MAN stands next to her, observing her with a tender smile. The GUIDE waits for people to get comfortable, points out places. Then, with a finger on his lips, he signals for silence and turns, like one more spectator, toward the characters who begin the action)] (3: 71-72; 72).

This scene affords both the spectators and the reader an important insight into the formulating of a spectacle within this play. Here the traditional set-up for a theatrical play—the entrance of the audience into the physical theater area, their accommodation into seats, the lighting of the scene—becomes a central part within the actual scene itself. Similarly, the guide, after assisting the spectators, blends into the audience and begins to view as a spectator the scene that is developing before them all. In this way, the line that is traditionally in place between the actors and spectators begins to smudge and fade away in favor of an intermingling between the two. When the spectators and the actors begin to come together, however, the play becomes something more than a simple story. What is going on onstage—a term used figuratively here since there is no traditional stage space—indicates something that can happen to the spectators also, a sensation that becomes all the more real when one of the "spectators" is kidnapped in Scene Six.

This scene quoted above continues with the man engaging in conversation with this same girl. It is obvious from their physical positions that the man is in control of the situation and, despite his tender smile, is about to exploit her. The girl's physical condition of being seated in a chair below the man and being completely wet suggest a violent beginning that the spectator has not seen—perhaps she has been subjected to the *submarino*—a practice of torture used in Argentina where the victim's head was submerged in water. Moreover, the characterization of her as a "muchacha [girl]" and of him as an "hombre [man]" emphasizes for the reader the difference of power and, thus, control of the situation that is symbolized in their inequality of height. He, the man, is older and established, towering above her both literally and figuratively while she, the girl, is below him and subject to his desires both physically and psychologically. The scene advances and the man censures the way she has been treated by some unnamed others before he entered the room, placing himself deceptively in the role of her protector. When it is obvious she feels cold, he gives her his jacket and in this way, tries to gain her respect and trust, an action that most importantly tries to seduce the audience.

This is an important moment in *Información* because Gambaro hints at the use of manipulation in the creation of a public event.[25] As we will see in a later quote, the man is fully aware of the presence of the spectators and thus attempts to present himself in a favorable light to them. This highlights how the torturer (because that is who this man is—a psychological torturer) performs a part that will deceive the public eye and manipulate what is true. The man's actions hide a sinister motive and point to a psychological violence that can be inferred from between the lines of his words and actions:

Hombre (*saca una pistola de la cintura y la limpia con un trapo*): ¡Ah! Es un servicio de mier... (*El Guía chista. El hombre lo mira fugazmente*) Sí (*Muestra el arma*) ¿Te gusta? Está descargada. (*Ella mira, no contesta. El hombre comienza a cargar el arma*) ¿Por qué tan triste? (*Señala al grupo*) No te pasará nada. Hay mucha gente. Nos miran. (*Guarda la pistola*) No quedás linda con el pelo mojado. Pero no es demasiado grave. (*Se inclina hacia ella, curioso*) Decíme, ¿te teñís el pelo? (*Sigue mirando*) Me estás mojando el saco. Perdón, es el único que tengo... (*Se lo saca suavemente, lo sacude y se lo pone. Con un estremecimiento*) Está húmedo... (*Señala la pistola*) ¿La querés?
Muchacha: No.
Hombre: ¡Te la dejo! Tengo otra. El saco no puedo, te lo juro.
Muchacha (*aparta la cabeza*): No.
Hombre (*subrepiciamente*): ¡Hablá más alto! ¡No oyen nada!
Guía: ¡Voce! ¡Voce!
Hombre: ¿Qué te decía? (*La muchacha no contesta*) Mirame. (*Ella obedece. Le tiende el arma*) ¡Tomá!
Muchacha: No... No quiero.

[*Man*: (He pulls a pistol from his belt and cleans it with a rag.) Ah! This department isn't worth shi... (The GUIDE says something. The MAN shoots him a quick look.) Right. (He shows her his weapon.) Do you like it? It isn't loaded. (She looks at it but doesn't answer. The MAN begins loading the gun.) Why so sad? (points to the group) Nothing will happen to you. There are lots of people. They're watching us. (puts the pistol back in his belt) You're not pretty with your hair all wet. But that's not too serious. (He leans toward her, curious.) Tell me, do you dye your hair? (still studying her) You're getting my jacket all wet. Sorry it's the only one I have... (He takes it gently, shakes it, and puts it on. With a shiver.) It's damp. (pointing to the pistol) Do you want it?
Girl: No.
Man: I'm leaving it for you. I have another. The jacket I can't, I swear to you.
Girl: (shaking her head) No.
Man: (surreptitiously) Speak up! They can't hear a thing!
Guide: Louder! Louder!

Man: What did I tell you? (The GIRL doesn't answer.) Look at me. (She obeys. He holds out the gun.) Take it!
Girl: No...I don't want to] (3: 72–73; 72–73).

This scene sets the tone for the progression of the entire work in that it outlines the roles of the guide, the audience and the stories themselves. After the guide has settled in his group in the previous quote, the action between the girl and the man begins to unfold. As they talk, or more accurately, as the man speaks to the terrified girl who listens, he takes out a pistol. When he is about to use inappropriate language, the guide quickly cuts in to reprimand him. This intervention by the guide highlights his/her role as in-between the actors and the audience; at times, the guide is the face of the play for the spectator while at others he/she is more identified with the audience. Similarly, the guide's interventions bring into question the whole play in that the spectators are viewing (and participating in) a product that does not seem to be finished or ready to be performed publicly.

This unfinished quality and the intermingling between actors and spectators that it demands force the spectators to reconsider their position and purpose in the representation. Should they intervene to put a stop to the girl's misery? Is the scene being put on for their benefit or, as the man says to the girl, are they the ones who are preventing anything more from happening? In this way, the spectators are forced to question their role in the play's development and, thus, reconsider their place in other situations outside the theater. This questioning is the ultimate goal of *Información para extranjeros* in that Gambaro aspires to revolutionize spectatorship both within and outside the theater through the horror of the scenes. Thus, Gambaro's play remembers Artaud's assertion that a spectator who witnesses a horrific violence onstage would be incapable of reproducing that violence offstage and drives the point further with the spectator's passive complicity.[26]

This scene between the man and the girl continues on to demonstrate a particularly harrowing experience of psychological violence. As he takes out and then loads his gun with bullets, the man attempts to engage the girl in small talk. He comments on her appearance and then takes away the jacket that he had offered her in order to keep warm earlier in the scene. His words try to indicate that she is the one overstepping social bounds by getting his jacket wet. In this simple action of taking away what had made him seem compassionate for a moment, the man reinforces the torturous route that the whole scene takes on. His words and actions place the girl in the role of offender, and, for both the girl and the spectator who understands the nature of the situation, makes his own actions all the more abhorrent. He wants to paint a picture where she is the one who is taking advantage of his chivalry by getting the jacket he so graciously lent her wet. Next, the man offers the girl the pistol; in effect, he tries to exchange his jacket for the gun, saying that he can give her the gun but not the jacket. With this move, the two begin to exchange words, he pushing her to take the gun and she refusing it. During this exchange, the man reminds the girl to speak up, remembering the audience that is listening to their words and the guide breaks in and seconds the man's reprimand. At this moment, the theatrical tension created in this intense scene is broken for all but the girl by underscoring the creation of a spectacle. The moment becomes a scene within a play and the horrific psychological violence that can be observed in the man's actions and words is divorced from the situation and turned into a performance for the spectators. This lessens the tension that has been created and makes it easier for the audience to relax by thinking that this is not real and not something to be worried about. Nevertheless, it simultaneously makes the spectators reflect upon the creation of this scene and the violence within it. Borrowing from Bertolt Brecht's writings on distancing, Gambaro creates a situation where the spectators become detached from the emotionally charged actions of the man and the girl to consider from a distance the situation before them.[27] However, this interruption of the scene runs the risk of trivializing the girl's ordeal and allowing the spectators' horror to dissipate. The audience can begin to question the validity of the girl's distress because it can be seen as unreal and performed. Nevertheless, it is my opinion that Gambaro creates this distance to once again underline the role of performance in both this scene and the quotidian episodes. Gambaro allows her audience the space in which they can both acknowledge the creation of a spectacle and reflect on the horror of its reality.

In the second half of this same scene between the man and the girl, the man's taunting of the girl continues and he finally leaves her alone with the loaded gun, though not before he plants a suggestion of what she can do with the pistol. In this part of the scene, the audience steps out of the action and the inhumanity that is shown by the man takes on a more sinister air in that one can forget that this is a play:

Hombre: [...] Tomá, agarrala. La dejo acá, en el suelo. No tenés más que inclinarte.
Muchacha: ¿Para qué? No quiero...inclinarme, no quiero...nada.
Hombre: El corazón y la frente...Son seguros. Digo, para no sufrir...
Muchacha: No...
Hombre (Le acaricia la mejilla): Claro que no. ¡Hay un sol afuera! Raja las piedras. ¿Así que no tenés novio? ¿Y entonces...? (*Se encamina hacia la puerta. Se vuelve. Sonríe*) ¡Voy a decir que calienten el agua! (*Sale*) (*La muchacha mira la pistola en el suelo, temblorosamente se inclina, tiende la mano. Se paraliza en el gesto*).

> [*Man*: [...] Take it! I'm leaving it here, on the floor. All you have to do is lean down.
> *Girl*: For what? I don't want...to lean down, I don't want...anything.
> *Man*: The heart and the forehead...are sure. I mean, so you don't suffer...
> *Girl*: No...
> *Man*: (caresses her cheek) Of course, no. There's a sun outside. It's hot as hell. So you don't have a boyfriend? Well then...? (He goes toward the door. Turns. Smiles.) I'm going to tell them to heat the water! (He goes out. The GIRL looks at the pistol, leans down, trembling, stretches her hand. Freezes in the act)] (3: 73–74; 74).

The man leaves the pistol on the floor for the girl and her response shows a hesitation. Picking up on this hesitation, the man plants an idea with the girl, suggesting that the heart and the forehead are the most secure places where a shot will cause a death without suffering. He then leaves the room, allowing the girl to use his suggestion on herself. Hidden within his words is also the idea of her using the gun to kill someone else, perhaps the man and, in theory, liberate herself from the suffering that she has to undergo at his hands. Nevertheless, he does not worry that this will come to pass since he turns his back on her and leaves the room, demonstrating how confident he is in his power and in the fear he has instilled in her. His final words break the previous connection he had attempted to make and confirm his role as one of her torturers. The girl's actions after the man's exit—moving to take the gun—reveal the depth of her fear and helplessness which were inspired by the man's actions and words and what she knows will come next.

Inherent in the Man's invitation to the Girl to use the gun on herself, we see a more sinister interpretation of his actions. Rather than kill her himself, he is offering a way to kill herself; in effect, he is shifting the blame for the action of ending a life from himself to the Girl, making it her own choice and her crime rather than his own. In this way, his offering becomes even more brutal and disturbing in that he wants to frame her destruction in the light of a choice that she made. By highlighting this attempted shift in responsibility, Gambaro encourages the spectators to understand the manipulations of choice present in the act of torture.

In the context of the play, this scene does not only hope to deter future actions of violence through the stage, as Artaud suggests, but also, by inviting the spectators to carefully consider the material present in the Brechtian way, wants to initiate a dialogue on the subject of torture itself. Here we are reminded of Foucault's considerations on public torture. Whereas before the eighteenth century, criminals were punished in public to serve as an example and a spectacle, this changed at the end of the eighteenth and the beginning of the nineteenth centuries.[28] Punishment became a hidden, shameful act. Additionally, there was a move to eliminate pain in executions, the introduction of the guillotine during the French Revolution being the result. These two points—the shame of punishment and the elimination of pain—are two important elements of the scene between the girl and the man. Though there is an audience, this conversation takes place privately. It is not open for public consumption; the spectators have paid to enter into the room. Like all acts of torture conducted in Argentina in the 1970s, this one is conducted in a dark and private room where someone controls who enters, leaves and how.

Furthermore, when he takes back his coat, the man's words seem to want to shame the girl for what she has done: "Me estás mojando el saco. Perdón, es el único que tengo... (*Se lo saca suavemente, lo sacude y se lo pone. Con un estremecimiento*) Está húmedo... [You're getting my jacket all wet. Sorry it's the only one I have... (He takes it gently, shakes it, and puts it on. With a shiver.) It's damp.]" His actions are gentle and he expresses a sadness that she has wet his jacket, giving the impression that *she* has overstepped the bounds, not him. He is ashamed and saddened by her actions. It is at this point that we witness the move towards execution, an execution that, if the man has his way, will be swift and painless. He leaves her the gun, suggesting she aim it at her heart or her forehead "para no sufrir... [so you don't suffer]" This illustrates what Foucault calls "an execution that affects life rather than the body."[29] The emphasis is on ending the life rather than prolonging the pain. However, this is also a mocking action in that everything, every movement, every word, every gesture is painful for the girl. The psychological effects of the conversation, of being in the room, of being half-naked and wet all add up to an imprisonment that is pain itself. Just as Foucault pointed out that, though pain should be avoided in executions, imprisonment still consisted of pain, here we see the meshing of the two and how cruel this combination of pain and non-pain can be.[30] Gambaro invites the spectators and readers to consider this double-speak and the long reaches of torture. Remembering the title of the play, this scene is intended to give us information about Argentina, but also about the painful effects of imprisonment and execution.

The title of the play—*Información para extranjeros* [*Information for Foreigners*]—relates directly with the argument that is central to Gambaro's intentions. *Información para extranjeros* indicates that the play is intended for foreigners, people that would not be able to understand the context and argument without some sort of explanation. The play, from the title, implies that it will provide the background information that will aid in understanding. Nevertheless, when the explanations are given, as we see, they do not reveal any information that goes beyond the context or argument presented. With this, Gambaro hints that all of the spectators that view or read her play are not equipped with the proper understanding to comprehend the situation before them. Instead, all of the spectators are foreigners to the situations before them and need to have the information explained, as can be seen when the guide says "Explicación: para extranjeros [Explanation: For Foreigners]." Gambaro underlines the fact that no one is equipped with the knowledge or the ability to comprehend what is happening. This can be seen in the following scene where the guide takes a moment to explain a reference.

The group and the guide enter a room where a woman in a white dress is rocking a doll. A man sits at her feet, looking up at the two. This scene is characterized as crudely acted by both Gambaro's words in the stage directions and the designations of the dress and characters. The guide expresses satisfaction at what seems to be a more "normal" scene as the woman begins to sing a lullaby to the doll. With this the guide turns to his group with the following explanation of a kidnapping:

> *Guía*: ¡Qué cuadro! (*A su grupo*) Acomódense. ¿Ven bien? Señora… (*ubica a una mujer. Luego, con voz seca y rápida*) Explicación: para extranjeros. Siete de la tarde del miércoles 16 de diciembre de 1979. Néstor Martins, abogado, defensor de presos políticos y gremiales, conversa con su cliente Nildo Zenteno. Se despiden en la calle. Seis hombres rodean a Martins, lo introducen a golpes en un Peugeot blanco. Nildo Zenteno se abalanza, consigue liberar fugazmente al abogado. Un golpe de karate en la nuca lo derrumbe también a él. El coche parte velozmente. Lo escolta un Chevrolet negro. Este coche ha salido cerca de la playa de estacionamiento de la Policía Federal. Desaparecidos. (*De los diarios*) Néstor Martins, 33 años. Nildo Zenteno, 37.
>
> [*Guide*: What a picture! (to his group) Make yourselves comfortable. Can you see? Madam… (helps her get comfortable. Then, rapidly, drily.) Explanation: For Foreigners. Seven p.m., Wednesday, December 16, 1970. Nestor Martins, attorney, defender of political prisoners and trade unions, consults with his client Nildo Zenteno. They take leave of one another in the street. Six men surround Martins, violently force him into a white Peugeot. Nildo Zenteno rushes back, manages momentarily to free the lawyer. A karate chop to the back of his neck brings Zenteno down as well. The car speeds off. A black Chevrolet escorts it. That car had pulled out of a nearby parking lot of the Federal Police. *Desaparecidos*. (from newspaper) Nestor Martins, thirty-three. Nildo Zenteno, thirty-seven] (5: 84–85; 85).

What is most interesting here is the explanation that sets up the scene as well as the juxtaposition between the visual part (the woman, the man and the doll) and the oral explanation that the guide provides. First, the audience participates in the creation of the scene that the guide is managing and, thus, becomes a part of the argument that is unfolding. The guide's role as an in-between underlines the necessity for a role that can negotiate the various spaces that are being juggled. None of these spaces can be understood on their own but need a guide that can act as a bridge—the spectators (and the readers) are the foreigners from the title that need a cultural and linguistic translation. This is seen in the juxtaposition of the woman singing a lullaby and the guide's story of a brutal kidnapping, which blurs the lines of theater. Rather than a correspondence between the two, the visual spectacle and the aural representation here clash against one another. The spectators, who have already been connected within the scene's account, become confused between the represented action and their own personal responses. How can this story be true when its background is that of a woman singing a lullaby to a child? At the same time, the child is a doll and reveals the artificiality in the creation of an event. What is real, then—what we see or what we hear? Gambaro aims to question not only the stories that have been reported as fact but also the creation of these very "facts." In this way, questioning, even questioning of the very play, is at the heart of *Información para extranjeros*.

This blurred line between reality and performance that is seen in the representation of actual episodes of torture continues in Scene seventeen. Here, there is another explanation directed at foreigners and the role of metatheater takes center stage in Gambaro's use of a rehearsal of Shakespeare's play *Othello*. The setting begins in the previous scene when the guide tries to enter one room with his/her group and is denied entry because "Estamos ensayando [We're rehearsing]" (16: 116; 118). The guide remembers that there is another entrance and the group of spectators accompanies him there and silently enters the room. They observe from behind screens a rehearsal of the final scene of *Othello*—when Desdemona has been killed and Iago is being confronted for his role in condemning Desdemona by both his wife Emilia and Othello himself. While the scene unfolds, a policeman in Elizabethan dress enters, playing a role within the scene that is not designated by the original script. Actor 1, who is playing the role of Iago, questions how the policeman entered and what he is doing while the actresses, in character, poke fun at the intrusion. All the while, the policemen begin to accuse the actors of murder and the scene takes on a much more sinister air that is compounded by the outcome:

> *Actor 1*: ¿Quién le dijo a éste que entrara?
> *Policía (actuando)*: ¡A mí, guardias! ¡A mí!
> *Actor 1*: ¡Andá a actuar a otro lado! ¿Quién te llamó? ¡Salí de acá!
> *Policía*: No tienes armas, y a la fuerza habrás de someterte. ¡Están muertas!

Actriz 1 (*se burla*): ¡Muerta soy!
Actriz 2: Sauce, sauce, sauce. ¡Moro, era casta! ¡Te amaba, moro cruel!
Actor 1: ¡Pará! (*al Policía*) ¿Te vas o no?
Policía: ¡Levantar vuestra espada contra una mujer!
Actor 2: ¿Qué dice?
[...]
Policía: ¡Guardias, a mí!
(*Entra otro policía, vestido del mismo modo*)
Policía 2: ¿Qué ocurre, mi señor?
Policía 1 (*saca del bolsillo una bolsita y la muestra*): ¡Trotyl! ¡Y ellas están muertas! ¡Señores, a mí! ¡Oh, pernicioso miserable!
Policía 2 (*arrinconando a los actores con su espada*): Vamos, ¡o los achuro!
(*Las actrices lanzan una risa incómoda*)
Policía 1: ¡Llévenlas también a ellas, por reírse a destiempo! (*Con tono de drama*) A vos, señor gobernador, incumbe la sentencia de este infernal malvado. Y de éste. Fijad el tiempo, el lugar, el suplicio. ¡Oh, que sea terrible! Yo voy a embarcarme inmediatamente y a llevar al Estado, con un corazón doloroso, el relato de este doloroso acontecimiento. (*Saca un arma del bolsillo, obliga a salir a los actores*).

[Actor #1: Who told this guy to come in?
Policeman #1: (acting, calling his men) Over here, men. Here!
Actor #1: Go act for the other side. Who called you. Get out of here!
Policeman #1: Thou hast no weapon, and perforce must suffer. They are dead.
Actress #1: (joking) I am dead!
Actress #2: (sings) Willow, willow, willow.
Moor, she was chaste. She loved thee, cruel Moor!
Actor #1: Stop! (to the Policeman) Will you beat it!
Policeman #1: To raise your sword against a woman!
Actor #2: What are you talking about?
[...]
Policeman #1: Officers, come here! (Another POLICEMAN enters, dressed in the same style.)
Policeman #2: What's happening, sir?
Policeman #1: (He shows him the vial he's just taken from his own pocket.) Trotyl! And the women are dead! Oh my! O thou pernicious caitiff!
Policeman #2: (with his sword, rounds up the ACTORS, who move into a corner) Move it, or I'll take a slice! (The ACTRESSES let out an inappropriate laugh.)
Policeman #1: Take them, too, for having laughed at the wrong time! (in a dramatic voice)
To you, Lord Governor,
Remains the censure of this hellish villain,
The time, the place, the torture, O, enforce it!
Myself will straight aboard, and to the state
This heavy act with heavy heart relate] (17: 118–119; 119–121).

With the entrance of the police officers in Elizabethan dress, this scene begins to double as both a ridiculous and sinister interpretation of Shakespeare and, as we will find out, of real-life events. The actors' reactions are important: Actor 1 gets angry at the intrusion and Actor 2 is confused, whereas the actresses confront the situation with humor, thinking that it must be a joke. Nevertheless, the confusion for both the actors of the Shakespearean scene and the spectators of Gambaro's play performs a vital part in the purpose of the acting. The Shakespearean actors, much like the audience, do not understand what is happening, given that the police enter into their theatrical space judging a fictional act. The actor is accused of killing two women, an act that has only taken place within the imagination of the characters rehearsing. The accusation is doubly ridiculous because the action is not only fictional but a private rehearsal. When the actors try to end the policeman's farcical behavior, they are instead accused of more illegal activity in possessing the explosive trotyl and carried away. The lines between reality and spectacle become blurred within the representation of *Othello* which takes place within the representation of *Información para extranjeros*. This circularity of representation creates a downward spiral that leaves the spectators (and there are multiple layers of spectators here—the actresses of *Othello*, the guide in *Información*, the audience of *Información*, the readers of *Información*) confused as to what is "real" and what is "fiction." Gambaro's intentions, then, spill over into other areas where this line between reality and fiction can erase itself and both become confused into one. Similarly, the multiple layers of spectatorship underscore the creation of the various spectacles within Gambaro's play. Fiction and reality are confused in the play in order to highlight the confusion and chaos of fiction and reality that exist in the temporal and historical context of 1970s Argentina.

Ending the scene above, the guide returns to his role in-between to try to explain the events that have occurred. By acknowledging the confusion that has been presented to them, there is another explanation intended for foreigners based on events and reports from the moment. Nevertheless, in this explanation, the ferocity of the events informs the guide's words and the commentary becomes an editorial before it can attempt to explain the confusion:

> Guía (a su grupo): Un poco confuso el desarrollo, ¿no? Para que entiendan. (*Pasa al espacio iluminado. Con voz profesional, seca y rápida*) Explicación: para extranjeros. (*Feroz y grosero*) ¿Alguien necesita realmente una explicación? Si ustedes se las dan de actores, se meten en un conventillo y aúllan como perros, la gente se asustará. Si no tienen plata, la gente se asustará más. ¿Por qué gritan? ¿Qué pretenden? Cuando nadie puede

abrir la boca, ¿por qué se va a gritar gratuitamente? (*Espera una respuesta que obviamente no recibe*) ¡Allá va! (*Retoma su tono profesional*) 6 de agosto de 1971. La policía irrumpe en una casa antigua, con muchas habitaciones como ésta, en la ciudad de Santa Fe. En una de las habitaciones se descubrió 800 gramos de trotyl. Dicen. Detenidos un periodista y tres integrantes del Grupo de Teatro 67. Trasladados a Buenos Aires bajo sospecha de desarrollar acción subversiva. El Fiscal aconsejó la absolución por aplicación del beneficio de la duda. Absueltos el 24 de mayo de 1972. (*Cambia de tono*) Pocos los llamados, muchos los elegidos. Nueve meses de gayola. Quedaron a la miseria. Y bueno, ¡así es la vida! (*Sale del espacio iluminado, se une a su grupo*) ¡Esperen! ¡La función continúa!

[*Guide*: (to his group) A bit confusing, the way that happened, don't you think? So you understand. (He walks into the light. In a professional, dry and rapid voice.) Explanation: For Foreigners. (fierce and rude) Does anyone really need an explanation? If you want to act like actors, just go into a tenement and howl like dogs, throw a good scare into people. If you don't have money, people will be even more afraid. Why scream? Why pretend? When no one can open his mouth, why would anyone scream gratuitously? (He waits for a response, which he doesn't get.) Okay, then! (resumes his professional tone) August 6, 1971. The police burst into an old house with many rooms, like this one, in the city of Santa Fe. In one of the rooms they find eight hundred grams of trotyl. They say. One journalist and three members of the Grupo 67 theater are arrested. They're taken to Buenos Aires on suspicion of subversive actions. The district attorney recommended they be absolved on the benefit of doubt. They were absolved May 24, 1972. (change of tone) Few are called, many are chosen. Nine months in the cage. In misery. Well, that's life! (He leaves the illuminated space, goes back to his group.) Wait! The show goes on!] (17: 119; 121).

This is another explanation that tries to link the material presented with actual violent events that take place outside of the theatrical walls and, thus, attempt to call the audience's attention to the country's atrocities. This combination of a verbal explanation and the visual reconstruction juxtapose the elements that define the theatrical experience: the marriage between a written text and a live performance of that text. Here though, the two don't necessarily correspond but attempt to work together in order to convey the reality of the situation. What is perhaps most disturbing about this scene is how the guide indifferently returns to his duties of explaining and leads the group with the same perceived level of zeal as he had before, despite the emotion that he had shown when faced with the Othello scene. With his flippant "Así es la vida [That's life]," he highlights the reaction that many spectators and readers have before the atrocities that take place in the world. His reaction shows the ability of humanity to ignore human suffering in order to carry on in the face of violence.

In this scene we see a certain deterioration from the order of earlier scenes such as the Milgram experiment and the torture of the girl with the wet hair, an important aspect that can be observed in the play's progression and then in the final scene. Here there is a descent in the written text that starts by portraying a certain orderly structure that, while brutal, testifies to the maintenance of a political or social connection. While the actions of the Man toward the Girl in Scene 3 are reprehensible, there is an organization and a purpose to his actions. However, as the written script moves on to its culmination in the final scene, scenes begin to lose this order as we can observe in the Othello scene described above where none of the actors understands what is happening. While in a performance of this play, the presentation of the scenes would depend upon the guides' routes, it is important to understand this disintegration of organization and order and how this reflects the immediate political and social context of the play. This is of particular interest in the final scene that will be analyzed at the end of this chapter.

The role of meta-theater permeating the *Othello* scene is central to Gambaro's goals in *Información para extranjeros*. As seen through her innovation of the role of the spectator, Gambaro seeks to force the theatrical experience to reconsider itself and its goal. Here the action focuses on a theater group and their rehearsal of *Othello*, centering the spotlight on the very issues that would pertain most closely to the material that the audience is viewing. This close examination of acting can be seen throughout the play in various ways. Primarily, the guide continuously refers to the level of acting in the scenes before them and the play's status as a spectacle, as can be seen in the quote below from scene five where the guide's commentary cuts through what he/she sees as a poor representation:

> *Guía*: ¡Qué mal actuado! Perdonen. Mejor busquemos otra cosa. (*Empuja a la gente hacia la puerta*) No todo el espectáculo es así. Espero.
> *Madre* (*con la misma voz de bebé idiota*): ¿Los castigaron mucho, papá?
> *Padre*: ¡Nunca más se supo!
> *Los dos*: ¡Je, je, je!
> *Guía (precipitadamente)*: ¡Vamos! ¡Vamos, señores! Por lo menos, necesitan un mes más de ensayos. ¡Qué bestias!
>
> [*Guide*: What horrible acting. So sorry. Let's look for something else. (He pushes the people toward the door.) The whole show's not like this. I hope.
> *Mother*: (same voice of a stupid baby) Did they punish them a lot, Daddy?
> *Father*: Nothing more was ever known!
> *Mother and Father*: Yea, yea, yea!

Guide: (cutting in) Let's go. Let's go, gentlemen. They need at least another month of rehearsal. What dunces!] (5: 86; 87).

The guide's commentary on the scene's acting once again reminds the spectators that they are seeing a representation and requires them to consider their role in and contribution to this representation. This self-conscious reflection that is inspired by the play's content and acting is a central element of its purposes in a larger picture. The objective is to force the spectators to be aware of their own role in the spectacle through their reactions to the content presented. As Postma observes in reference to the Othello scene quoted earlier: "The guide manipulates the experience of the spectator in the same way that the playwright manipulates the lives of characters; and the spectators, deprived of the security of their usual distance as audience, become living witnesses to the recreation of a stage space-within-a-space where the frontiers between stage and life have become blurred."[31] In this way, the revolutionary aim of the play is to bring the spectators' (and readers') awareness to their own complicity in the violent acts unfolding before them.

The final scene of *Información*, the only one that Gambaro instructs that all of the groups should witness together, is a pivotal moment that culminates the innovation of space, spectator, and the questioning of the spectator's complicity in the violence in the play and the outside context. In this scene, the guides lead their groups to an open space where the other groups also converge. In this open area, there is music playing and prostitutes dancing. Subsequently, a man enters with two other prostitutes followed by four men with a blindfolded prisoner who all of the men begin to taunt. They twirl him around, making him lose his orientation. At this point, one of the prostitutes reaches out to take off his blindfold and she is immediately rebuked by one of the men, instantly thwarting any possibility of intervention and setting an example for anyone else who might wish to stop the action. Following this, the men strip the prisoner, and begin to fight like children over control of the situation and, finally, they all strike the prisoner and drag him behind a screen, where they place him on top of a table. All but two men come out from behind the screen; they turn the music up and force the prostitutes to sing and dance. Behind the screen, the two men begin to torture the prisoner, showing just the outline of the horrific scene. At this moment, with these two contrasting scenes, there is a juxtaposition between the jovial party-like atmosphere and the outline of the prisoner being tortured, which makes the spectator associate the two actions and *see* what is happening right next door. Gambaro forces the spectator and reader to question his/her own complicity within the spectacle. In this way, the spectator is not only the members of the group there but also the readers of the play.

The conclusion of the scene, and the play, comes quickly after this moment when the lights go out for a moment. When they come back on, the screen has been removed to reveal the prisoner's, now murdered, body. With that, he gets up, puts on his clothing and walks out. And with this, the play ends. The guides send the spectators away repeating the sentence: "¿Quién dijo alguna vez: hasta aquí el hombre, hasta aquí, no? [Who once said: here the ken // of men and women // here the bounds?]" (20: 128; 130). This meta-theatrical ending where the actor gets up at the end of the scene and walks offstage highlights the construction of the spectacle and makes the spectator stop and think about what is happening rather than become caught up in the emotion of the representation. Gambaro borrows from Brecht's thoughts on distancing in order to interrupt the emotional horror of the moment and to instead force the spectators to interrogate the scene that has unfolded before them and their contribution to its development. Similarly, the question that is repeated at the end of the play ("¿Quién dijo alguna vez: hasta aquí el hombre, hasta aquí, no?") again explores the limits of horror and atrocity that we all believe we hold. Instead, Gambaro places in doubt this line and makes the spectators question how far each one would go. This final scene, then, unites the audience both physically and emotionally against the practice of torture and the use of violence as a psychological weapon. From the action portrayed, the spectators and the readers are obligated to interrogate how their own actions may allow the continued use of violence. The innovation of space and spectatorship, then, are employed to question and provoke change in the audience.

Información para extranjeros by Griselda Gambaro is a tour-de-force that highlights the spectator's role in the creation of a play as well as our own passive contribution to violence and torture. Gambaro's central project is to wake up the spectators, literally taking them out of their seats and including them in the violent representations that she elaborates. Gambaro belongs to the utopian tradition of Latin American theater in which the means of production are at the disposal of the ends of the representation. That is, if the project here is to wake up the spectators and force them to see their own complicity with violence and torture, Gambaro will use radical means to accomplish this task. *Información* is perhaps the most radical of the four plays studied here. The play reveals a singular commitment to the moment in which she writes and an immediacy to both her historical and temporal contexts. Gambaro approaches her material directly and does not allow her spectators to confuse what she is saying. The spectator, and for *Información*, the reader, becomes a catalyst that will interrogate the relationship between actions that are represented onstage as well as those that are taking place outside of the theater. It is through these conversations provoked by the written text and its possible representations that *Información para extranjeros* aims to enter the debate on the use of torture, a debate to which we need to return in the current climate. The theater *is* the site of debate and innovation on the current events of the time. Gambaro's theater, then, becomes a possibility for an alternative space of social change that innovates through spectacle, a possibility that was shattered three years later when the *Proceso de Reorganización Nacional* arrived to power in 1976 and caused the exile of numerous playwrights, Gambaro included.

Conclusion

Transforming Spectacles

The sacrifice serves to protect the entire community from its own violence.
René Girard, Violence and the Sacred

This project studied how violence and violent acts were portrayed from 1968 to 1974 in Cuba and Argentina in four canonical plays by Virgilio Piñera, Abelardo Estorino, Eduardo Pavlovsky, and Griselda Gambaro. These years were defined by official violence that originated with the state against its people and by an extra-official violence that came from the nongoverning groups as a way to contest conditions or provoke change. The violence defined movements on both the left and the right of the political spectrum and came to be a fundamental part of global events as we see in such diverse settings as Paris and Havana. This study showed how theater answered this real, quotidian violence by placing it onstage in order to highlight the central role it has in everyday lives and the depth of its control over our lives. As shown here, these plays do not necessarily advocate an end to violence or didactically teach the spectators how to eradicate violence from their lives. Similarly, this project does not propose a list of plays that look at the topic of violence, but studies how violence has been understood through the stage. Piñera, Estorino, Pavlovsky, and Gambaro underline the presence of violence in the definitions of everyday life in an attempt to understand its role. While this can appear to be a simple act, the varying levels of censorship that the plays have experienced testify to the radical interventions that this highlighting of violence proposes. These plays written within a global and local framework of violence move the communities and the spectators towards a comprehension of the role of violence within our everyday relationships, our constructions of history, our identities, and our concept of our surroundings. It is in this central role of violence within our lives that these plays radicalize the stage to redefine our own relationship with violence in order to move beyond the present to a more utopian future.

The daily presence of violence in everyday lives can be seen in these plays, though not all of them refer directly to it. Piñera's *Dos viejos pánicos* portrays Tota and Tabo's fearful life in their own house, where violence manifests itself as an interpersonal element of their relationship with one another. Violence allows the couple to hide their fear from each other and paradoxically becomes a tool that permits them to hide from reality. Estorino's *La dolorosa historia* uncovers the role of violence in the construction of Cuban history and, consequently, in Cuban political events. Violence here is an inherent part of historical interpretation that simultaneously allows Estorino to allude to his own violent moment in Cuban history. Pavlovsky's *La mueca* places the play's action in a bourgeois couple's apartment and foregrounds the violence inherent in constructions of identity of class, gender, and sexuality. Violence becomes an instrument used to create and inspire one's own and others' identity forced from without. Gambaro's *Información para extranjeros* violently transforms normative theatrical structures by dictating that her play should unfold in a house rather than in a theater and that the spectators should be guided through the rooms as the play progresses. These violent and radical gestures to the conventional notion of theater in Gambaro are multiplied by the violent acts portrayed in the play's scenes. Thus, Gambaro creates a piece that engages its spectators and questions the national context in which it was written. Within the spaces represented by these playwrights, violence cannot be denied as part of an everyday existence. These plays allow us, then, to understand the presence of violence in our lives so that we can control it rather than allow it to control us. Nevertheless, at the same time, violence is revealed to be an unpredictable, powerful force that can be contradictory. The playwrights endeavor to show the multiplicity and complexity inherent in violence and the need to comprehend its role within lives and communities in order to not be dominated by violence.

The four plays studied here exemplify the kinds of responses that theater offered to the political context of the late 1960s and early 1970s. They all focus on how violence manifests itself within a community, and how it should be destroyed or kept in check. The fascination with violence in these four plays is undoubtedly a legacy from earlier theoreticians' considerations of its inevitable role within our lives. Antonin Artaud believed that the use of violence in theater would make it impossible for spectators to recreate violent acts outside of a representation. For René Girard, violence was a necessary part of primitive societies that could be expelled through ritual sacrifice. Elaine Scarry and Michel Foucault's writings highlight the role of spectacality in violence and, thus, connect it with theater and representation. Piñera, Estorino, Pavlovsky, and Gambaro use the inherent relationship in theater between the word and the image to stress their own visions of the violent contexts in which they interact in order to create a spectacle that offers an alternative response to violence. Their vision is, in many ways, utopian—for they believe that an end to violence can only come about by means of an act of consciousness of its place in our lives.

Piñera, Estorino, Pavlovsky, and Gambaro present violence in their plays in intersection with other aspects of life. As we saw with the analysis of the Padilla Affair, violence and spectacle formed a pivotal part of Latin American communities of the 1960s and 1970s. The Padilla Affair and the events that surrounded it constructed a representation that both sides attempted to manipulate and control. On one side, the revolutionary officials tried to muzzle Padilla's criticism and to orchestrate a public confession that would discredit the poet and silence future critics through fear. Padilla, on the other hand, used the genre of the public confession to ironically expose his sins and, in the process, uncover the strong-arm techniques and the repressive measures of the Revolutionary government. He presented himself as the penitent intellectual, playing the part that the Revolution demanded of him. The Revolution, in turn, demanded the presence of other writers and artists and succeeded in sending a message of fear. However, neither side can be said to have won. Padilla suffered torture and eventually exile and the intellectual community in Cuba endured years of silence and censorship. The Revolution, for its part, revealed its repressive tactics for an international audience. Looking further afield, this Affair unveils the central role that the creation of spectacle held, an importance that these playwrights used in their theater.

In this study, I have used the topics of fear, history, identity, and spectatorship as starting points that allowed me to understand how violence is manifested in these works. In the course of my analysis, I came to the conclusion that violence covers or uncovers something that specifically relates to each of the characters and the situations presented. That is, the playwrights speak to the particular context in which they write. The quote from René Girard that opened this chapter maintains that sacrifice protects its community. In these plays, though, the violence on stage aims to open a dialogue and make visible the central role that violence was already playing in the surrounding communities, in order to unmask the violent tendencies that were prevalent. Theater wanted to reveal this role to its spectators to create a situation where the audience was aware of its own actions and the consequences inherent in them. This would, in turn, create a community where violent acts would be understood before their destruction could be irreversible, and could be stopped. Theater here becomes a utopian tool to intervene against violence and its manifestations in a violent society.

In *Dos viejos pánicos*, the violent gestures that the two characters exchange hide the deep fear that they have of all that exists outside of their small world. The violent acts that bind them allow them to repress their fears. This play, chronologically the first one written of those studied in this book, in fact opens a violent period in Latin American history. Piñera underscores that violent interpersonal relationships reflect the political context found in Cuba, specifically—and in Latin America, generally—during these years. In *Información para extranjeros*, Gambaro directly alludes to the historical context in which she writes. Instead of focusing on the "closed" circuit of interpersonal violence, she does violence to the notion of theater itself, in order to raise her spectators' consciousness. Gambaro's is the most radical play studied here, for she revolutionizes theater in order to establish direct connections between her own material and what is taking place outside of the space where the theatrical piece is being represented. For Gambaro, violence is the product of an insane world, and it needs to be brought directly into the space of representation by any means possible.

Abelardo Estorino also radically and violently alters notions of representation in *La dolorosa historia*. Here, violence is a definite presence in the creation of history. By refocusing his spectators' eyes on the past, Estorino sheds light on the violent context of the early 1970s in Cuba and questions the effects this continuing violence will have in the future. In Pavlovsky's *La mueca*, the only of these four plays to premiere at the moment of writing, violent gestures are used as a way for one character to construct his/her own identity while forcing another into a space of subjugation. Pavlovsky, then, forces his spectators to confront how their own definitions of Self and Other are the result of violent affirmations of identity.

Theater is one of the most distinctly pertinent genres from which to engage an audience on the future of a community. It is collaborative and immediate; it possesses a certain urgency, and it is theoretically performable anywhere. These qualities allow theater a space from which to connect with its spectators and contribute to a society's future. Piñera, Estorino, Pavlovsky, and Gambaro all engage their audiences around violence and violent acts. Their plays endeavor to understand the inevitable place of violence in our lives and the way that it affects relationships, in the creation of history, in the construction of identity, and theater itself. Violence for these four playwrights is an inescapable presence that needs to be acknowledged and recognized in order to be understood and controlled. These plays engage their audiences and their communities in ways that underscore the importance of the genre and allow theater to have a central role in their national contexts. Theater, in the words of Augusto Boal, is political and, thus, offers a pivotal, ever-changing dialogue on the place of violence in our world.

Notes

Preface   Understanding the Place of Theater in Spanish America

1. Adam Versényi, *Theatre in Latin America: Religion, Politics, and Culture from Cortés to the 1980s* (Cambridge: Cambridge University Press, 1993) 1–35.
2. Diana Taylor, *Theatre of Crisis: Drama and Politics in Latin America* (Kentucky: University Press of Kentucky, 1991) 2.
3. Versényi 59–62.
4. Jill Lane, *Blackface Cuba, 1840–1895* (Philadelphia: University of Pennsylvania Press, 2005) 15.
5. Versényi 72–78.
6. Sandra Cypess' essay on twentieth-century Spanish American theater in the *Cambridge History of Latin American Literature* should be consulted for more details on the theatrical production of this century. She examines the theater of the 1920s and 1930s through an exploration of three areas: Mexico, Puerto Rico, and Argentina and how the theater began to define itself autonomously in writing and theater groups. From this point, Cypess offers a broad view of the theater of the twentieth century through a presentation and examination of the major contributors and their works. This essay is, without a doubt, a much more complete overview of Spanish American theater of the twentieth century. Sandra Cypess, "Spanish American Theatre in the Twentieth Century," *The Cambridge History of Latin American Literature. Volume 2: The Twentieth Century*, ed. Roberto González Echevarría and Enrique Pupo-Walker (Cambridge: Cambridge University Press, 1996) 497–525.
7. Antonin Artaud, *El teatro y su doble*, Trad. Enrique Alonso y Francisco Abelenda (Barcelona: Edhasa, 1978) and *Theater and Its Double*, Trad. Mary Carolina Richards (New York: Grove Press, 1958). Martin Esslin, *The Theatre of the Absurd* (Woodstock, NY: Overlook Press, 1973).
8. Artaud 79.
9. George Woodyard, "The Theatre of the Absurd in Spanish America," *Comparative Drama* 111.3 (1969): 186.
10. Raquel Aguilú de Murphy, *Los textos dramáticos de Virgilio Piñera y el teatro del absurdo* (Madrid: Editorial Pliegos, 1989). Daniel Zalacaín, *El*

teatro absurdista hispanoamericano (Valencia, España: Ediciones Albatros Hispanófila, 1985).
11. Eleanor Jean Martin, "*Dos viejos pánicos*: A Political Interpretation of the Cuban Theater of the Absurd," *Revista/Review iberoamericana* 9 (1979): 50–56. Terry L. Palls, "El teatro del absurdo en Cuba: El compromiso artístico frente al compromiso político," *Latin American Theatre Review* 11.2 (1978): 25–32. Woodyard 183–192.
12. Fernando de Toro's comprehensive *Brecht en el teatro hispanoamericano* offers an in depth view of the Brechtian theory and theater and how it influenced and manifested itself in Spanish America. Fernando De Toro, *Brecht en el teatro hispanoamericano*, Buenos Aires: Editorial Galerna, 1987.
13. Bertolt Brecht, *Brecht on Theatre: The Development of an Aesthetic*, ed. and trans. John Willett (New York: Hill and Wang, 1964) 190.
14. Brecht 192.
15. John Fuegi, *The Essential Brecht* (Los Angeles: Hennessey and Ingalls, 1972) 15.
16. Fuegi 18.
17. Nora Eidelberg, *Teatro experimental hispanoamericano, 1960–1980: La realidad social como manipulación* (Minnesota: Institute for the Study of Ideologies and Literature, 1985) 1–11.
18. Jorge Castañeda's book *Utopia Unarmed* offers the best discussion of leftist initiatives in Latin America, their results and current states. Jorge G. Castañeda, *Utopia Unarmed: The Latin American Left after the Cold War* (New York: Alfred A. Knopf, 1993).
19. Augusto Boal, *Theatre of the Oppressed*, trans. Charles A. and Maria-Odilia Leal McBride (New York: Theater Communications Group, 1985) ix.
20. Boal ix.
21. Boal x.
22. Antonin Artaud, *Theater and Its Double*, trans. Mary Caroline Richards (New York: Grove Press, 1958) 81.
23. Diana Taylor and Roselyn Costantino, *Holy Terrors* (Durham, NC: Duke University Press, 2003) 11–13.
24. Taylor and Costantino 4.
25. Taylor and Costantino 4–5.
26. Huerta, Jorge, *Chicano Theater: Themes and Forms* (Tempe, Arizona: Bilingual Press/Editorial Bilingüe, 1982) 1.
27. Huerta 3.

Introduction Difficult Times: Considering Dramatic Violence

1. This idea of spectacularity refers to the creation or essence of a spectacle that is taking place outside of the theater. I am thinking of the Spanish word "espectacularidad," which suggests the construction and elements of

a spectacle. I will use spectacularity to refer to this concept throughout this project.
2. Bertolt Brecht, *Brecht on Theatre: The Development of an Aesthetic*, ed. and trans. John Willett (New York: Hill and Wang, 1964) 275.
3. Ernesto Che Guevara, "El hombre nuevo," *Ideas en torno de Latinoamérica*, Vol I (México: Universidad Nacional Autónoma de México, 1986) 322. Translation mine.
4. For Che, there was a clear connection between the masses and the vanguard that would bring about the revolution (Guevara 313).
5. José M. Fernández, "Teatro Experimental entrevista con Vicente Revuelta," *Conjunto* 3 (1964): 58–62. The Teatro Escambray in Cuba is another example of a theater group that attempted to bridge the gap between the people and the theater and can be seen as perhaps the most famous of many of these projects. Translation mine.
6. Michel Foucault, *Discipline and Punish: The Birth of the Prison*, trans. Alan Sheridan (New York: Random House, 1977) 8.
7. Foucault 14.
8. René Girard, *Violence and the Sacred*, trans. Patrick Gregory (Baltimore: Johns Hopkins University Press, 1977) 8.
9. Girard 10.
10. Girard 26.
11. Hannah Arendt, *On Violence* (New York: Harcourt, Brace & World, 1970) 54.
12. Arendt 80.
13. Frantz Fanon, *Wretched of the Earth*, trans. Constance Farrington (New York: Grove Press, 1963) 61.
14. Elaine Scarry, *The Body in Pain: The Making and Unmaking of the World* (New York: Oxford University Press, 1985) 4.
15. Emphasis from the original, Scarry 13.
16. This can be seen in Virgilio Piñera's *Una caja de zapatos vacía* (written in 1968, though not published until 1987), where an empty shoe box is in the feminine space of the tortured and Carlos occupies the role of torturer. He hits the shoe box repeatedly, even threatening it sexually for it represents something feminine.
17. Ariel Dorfman's *La muerte y la doncella* (1991) is a good example of this inversion.
18. Thanks to a grant from the University Fund for Internationalization from the Graduate School of Arts and Sciences of Emory University, I was able to engage in discussions in Havana, Cuba on theater of the 1960s and 1970s. From conversations with Norge Espinosa Mendoza and Abelardo Estorino, I was told of the rich tradition of reading theater from this time period. Both *Dos viejos pánicos* and *La dolorosa historia* benefited from this practice (Merlin H. Forster, "Games and Endgames in Virgilio Piñera's *Dos viejos pánicos*," *In Retrospect: Essays on Latin American Literature*, ed. Elizabeth S. Rogers and Timothy J. Rogers [South Carolina: Special Literary Publications., 1987] 105). *Información para extranjeros*, in turn, was

circulated as a manuscript when it was first written (Diana Taylor, "Theater and Terrorism: Griselda Gambaro's 'Information for Foreigners'," *Theatre Journal* 42.2 [1990]: 168).
19. Idelber Avelar, *The Letter of Violence: Essays on Narrative, Ethics, and Politics* (New York: Palgrave, 2004) 1.
20. Eric Hobsbawm, *The Age of Extremes: A History of the World, 1914–1991* (New York: Vintage-Random House, 1996) 22.
21. Hobsbawm 24.
22. Hobsbawm 26.
23. Thanks to a travel grant from the University of Florida's Latin American Collection, I was able to conduct extensive research at the University's library on the Padilla Affair and record how it unfolded in the contemporary journals and magazines.
24. *El caimán barbudo* opened an inquest about this since, as they say, Otero's novel had sold out within a week. *El caimán barbudo* 15 (1967).
25. As we can see with the passage of time, Cabrera Infante's novel was clearly superior. *Tres tristes tigres* continues to be read and discussed vigorously whereas Otero's *Pasión de Urbino* is not.
26. Manuel Díaz Martínez details the pressures that the Cuban jurists felt during this process in his essay "El caso Padilla: crimen y castigo (Recuerdos de un condenado)." *Inti: Revista de literatura hispánica* 46–47 (Otoño 1997– Primavera 1998): 157–166.
27. "Declaraciones de la UNEAC acerca de los premios otorgados a Heberto Padilla en Poesía y Antón Arrufat en Teatro," *Fuera del juego* de Heberto Padilla (Miami: Ediciones Universal, 1998), 115. Translation mine.
28. "Primera Carta de los intelectuales europeos y latinoamericanos a Fidel Castro," *Fuera del juego* de Heberto Padilla (Miami: Ediciones Universal, 1998), 123. Translation mine.
29. "Segunda carta de los intelectuales europeos y latinoamericanos a Fidel Castro," *Fuera del juego* de Heberto Padilla (Miami: Ediciones Universal, 1998), 160–1.
30. "Segunda carta de los intelectuales europeos y latinoamericanos a Fidel Castro" 160. Translation mine.
31. Fidel Castro, "Discurso de clausura," *Casa de las Américas* 9.65–66 (March– June 1971) 25.
32. Castro 27. Translation mine.
33. As quoted in Heberto Padilla, "Intervención en la Unión de Escritores y Artistas de Cuba, el martes 27 de abril de 1971," *Fuera del juego* (Miami: Ediciones Universal, 1998), 135. Translation mine.
34. Padilla, *Fuera del juego* 135. Translation mine.
35. Claudia Gilman discusses the role of journals in Latin America in the 1960s in the introduction to her seminal study *Entre la pluma y el fusil*. Claudia Gilman, *Entre la pluma y el fusil: debates y dilemas del escritor revolucionario en América Latina* (Buenos Aires: Siglo veintiuno editores, 2003) 22–26.
36. Lourdes Casal maintains that this was most likely José Antonio Portuondo, a well-known Revolutionary critic who would also preside over Padilla's

self-criticism at UNEAC headquarters. However, Roger Reed asserts that this was really Luis Pavón, the magazine's director, a statement that Ambrosio Fornet supports in 2006. Lourdes Casal, "Literature and Society." *Revolutionary Change in Cuba*, ed. Carmelo Mesa-Lago (Pittsburgh: University of Pittsburgh Press, 1971) 20; Roger Reed, *The Cultural Revolution in Cuba* (Geneva: Latin American Round Table, 1991) 106; Ambrosio Fornet, "El Quinquenio gris: Revisitando el término" (Havana 2006. http://www.criterios.es/pdf/fornetquinqueniogris.pdf) 9.
37. Leopoldo Ávila, "Las provocaciones de Padilla," *Verde Olivo* 9:45 (1968) 17–18. Translation mine.
38. Leopoldo Ávila, "Sobre algunas corrientes de la crítica y la literatura de Cuba," *La Gaceta de Cuba* Noviembre-Diciembre 1968, 3. Translation mine.
39. Heberto Padilla, *Provocaciones* (Madrid: Ediciones La Gota de Agua, 1973) p 41. Translation mine.
40. Reinaldo Arenas, "El caso y el ocaso de Padilla," *Fuera del juego* de Heberto Padilla (Miami: Ediciones Universal, 1998), 164.
41. Reed 117.
42. Reed 119–120.
43. Padilla, *Fuera del juego* 13–14; Heberto Padilla, *Sent off the Field: A Selection from the Poetry of Heberto Padilla*, trans. J. M. Cohen (Great Britain: Andre Deutsch, 1972).
44. Piñera's theater has been the object of many academic studies, most notably Rine Leal's 2002 prologue to Piñera's *Teatro completo*. The play *Dos viejos pánicos*, specifically, has received less attention, though essays by José Corrales, Kristin E. Shoaf, Merlin H. Forster, Hortensia Ruíz del Vizo, Leon F. Lyday, Eleanor Jean Martin, and Susana Howell stand out. José Corrales, "Los acosados, Tabo, Tota, Montes Huidobro y Piñera," *Círculo: Revista de cultura* 29 (2000):114–125. Kristin E. Shoaf, "Los ciclos herméticos y el miedo al miedo en *Dos viejos pánicos* de Vigilio Piñera," *RLA: Romance Languages Annual* 9 (1997): 689–92. Merlin H. Forster, "Games and Endgames in Virgilio Piñera's *Dos viejos pánicos*," *In Retrospect: Essays on Latin American Literature*, ed. Elizabeth S. Rogers and Timothy J. Rogers (South Carolina: Special Literary Publications, 1987). Hortensia Ruíz del Vizo, "Análisis del miedo en *Dos viejos pánicos* de Virgilio Piñera," *National Symposium on Hispanic Theatre*, ed. Adolfo M. Franco (Cedar Falls: University of Northern Iowa, 1985). Leon F. Lyday, "De rebelión a morbosidad: Juegos interpersonales en tres dramas hispanoamericanos," *Actas del sexto congreso internacional de hispanistas* (Toronto: University of Toronto, 1980). Eleanor Jean Martin, "*Dos viejos pánicos*: A Political Interpretation of the Cuban Theater of the Absurd," *Revista/Review iberoamericana* 9 (1979): 50–56. Susana Howell, "*Historias para ser contadas* y *Dos viejos pánicos*: Crítica y espectáculo," *Prismal/Cabral: Revista de literatura hispánica/Caderno afro-brasileiro asiático lusitano* 1 (1977): 17–24.
45. Virgilio Piñera, *Teatro completo*, ed. Rine Leal (La Habana: Editorial Letras Cubanas, 2002) 479. All translations are mine.
46. Piñera vi.

47. Though Estorino is considered one of the most important playwrights of the Cuban Revolution, there has been very little written on this play specifically. Jorge Febles' article on the use of verse in *La dolorosa historia del amor secreto de don José Jacinto Milanés* is the one exception. Jorge Febles, "Recontextualización poemática en *La dolorosa historia del amor secreto de don José Jacinto Milanés*," *Latin American Theatre Review* 31.2 (1998): 79–95.
48. Eduardo Pavlovsky is the author of numerous plays and the subject of many studies on theater. The essays by Matías Montes-Huidobro and Charles B. Driskell stand out as the most pertinent to this study. Matías Montes-Huidobro, "Psicoanálisis fílmico-dramático de *La mueca*," *Monographic Review/Revista monográfica* 7 (1991): 297–314. Charles B. Driskell, "Power, Myths and Aggression in Eduardo Pavlovsky's Theater," *Hispania: A Journal Devoted to the Teaching of Spanish and Portuguese* 65.4 (1982): 570–579.
49. Patrice Pavis, *Dictionary of the Theatre: Terms, Concepts, and Analysis*, trans. Melvin Carlson (Toronto: University of Toronto Press, 1988) 261.
50. Diana Taylor, "Opening Remarks," *Negotiating Performance: Gender, Sexuality, and Theatricality in Latin/o America*, ed. Diana Taylor and Juan Villegas (Durham, NC: Duke University Press, 1994) 11.
51. Griselda Gambaro's theater is the subject of hundreds of articles and numerous books. Some names that stand out as most important for the analysis of this play are Jason Cortés, Myriam Yvonne Jehenson, Mady Schutzman, John Fleming, Rosalea Postma, and Dick Gerdes. Jason Cortés, "La teatralización de la violencia y la complicidad del espectáculo de *Información para extranjeros* de Griselda Gambaro*,*" *Latin American Theatre Review* 35.1 (2001): 47–61. Myriam Yvonne Jehenson, "Staging Cultural Violence: Griselda Gambaro and Argentina's 'Dirty War'," *Mosaic: A Journal for the Interdisciplinary Study of Literature* 32.1 (1999): 85–104. Mady Schutzman, "Calculus, Clarivoyance, and Communitas," *Women and Performance: A Journal of Feminist Theory* 10 (1999): 117–33.John Fleming, "Argentina on Stage: Griselda Gambaro's *Information for Foreigners* and *Antígona furiosa*," *Selected Proceedings: Louisiana Conference on Hispanic Languages and Literatures*, ed. Joseph V. Ricapito (Baton Rouge: Louisiana State University Press, 1994). Rosalea Postma, "Space and Spectator in the Theatre of Griselda Gambaro: *Informacion para extranjeros*," *Latin American Theatre Review* 14.1 (1980): 35–45. Dick Gerdes, "Recent Argentine Vanguard Theatre: Gambaro's *Información para extranjeros*," *Latin American Theatre Review* 11.2 (1978): 11–16.
52. Griselda Gambaro, *Teatro 2. Dar la vuelta. Información para extranjeros. Puesta en claro. Sucede lo que pasa* (Buenos Aires: Ediciones de la Flor, 1987) 70. Marguerite Feitlowitz (Illinois: Northwestern University Press, 1992) 70. All text references in Spanish will come from the former and in English are from the translation by Marguerite Feitlowitz.
53. Gambaro 71; 71.
54. Gambaro 71; 71.

1 Who's Afraid of Virgilio Piñera?: Violence and Fear in *Dos viejos pánicos* (1968)

1. Marifeli Pérez-Stable, *The Cuban Revolution* (New York: Oxford University Press, 1993) 98–99.
2. Julio Matas, "Theater and Cinematography," *Revolutionary Change in Cuba*, ed. Carmelo Mesa-Lago (Pittsburgh: University of Pittsburgh Press, 1971) 432. Antoni Kapcia, *Havana: The Making of the Cuban Culture* (Oxford: Berg, 2005) 140.
3. Kapcia 140.
4. Cinema was another artistic genre that initially benefitted from the Revolution's desire to foment the arts, as outlined in Julio Matas' article. In March of 1959 the *Instituto cubano del arte e industria cinematográficos* (ICAIC [Cuban Institute of Cinema Arts]) was formed with Alfredo Guevara, a former student leader interested in the arts, as the head. Many of the important figures of Cuban cinema quickly joined the staff and the ICAIC confronted some revolutionary topics in its first few years. However, like theater, things began to change around 1961. Initially, this stemmed from a controversy around the film *P.M.* by Sabá Cabrera Infante and Orlando Jiménez. It portrayed a Saturday evening on the water in the capital where people were looking for entertainment, drinking and dancing. This was seen to be a frivolous topic and the ICAIC refused to allow the filmmakers space to screen the movie and confiscated it. This sparked a fierce debate involving *Casa de las Américas* and the final decision upheld that of the ICIAC. Cinema, like the other artistic genres, would need to confirm the ideals of the Revolution. Furthermore, production slowed around 1968 due to a shortage of film. This resulted in most of the movies made being documentaries that were favorable to the Revolutionary goals. Perhaps the most well-known movies from the 1960s are *La muerte de un burócrata* (1966), *Memorias del subdesarrollo* (1968) both from Tomás Gutiérrez Alea, and *Lucía* (1968) by Humberto Solás. Matas 438–442.
5. Abelardo Estorino. Personal interview. May 8, 2007.
6. Kapcia 141.
7. Kapcia 141.
8. Paul Christopher Smith, "Theatre and Political Criteria in Cuba: Casa de las Américas Awards, 1960–1983," *Cuban Studies/Estudios Cubanos* 14.1 (1984): 43–47.
9. Georgina M. Dopico Black's comprehensive essay on censorship in Cuba, "The Limits of Expression: Intellectual Freedom in Postrevolutionary Cuba," details three possibilities of censorship: "literature may be promoted, prohibited, or marginally tolerated by the official apparatus" (108). For each of these three options, Dopico Black offers a discussion of what this meant within the Cuban context and examples of how this was employed. This article is one of the most interesting and wide-ranging discussions of the employment of censorship in Cuba from the beginning of the Revolution until the fourth UNEAC Congress

in January of 1988. Antonio Benítez-Rojo responded to this reading of postrevolutionary Cuban censorship in "Comments on Georgina Dopico Black's 'The Limits of Expression:Intellectual Freedom in Postrevolutionary Cuba'" by expanding her views. Benítez-Rojo underlines the importance of the author, not just the work, within the decision to censor or not and questions Dopico Black's view that the 1988 UNEAC Congress demonstrated an opening in intellectual freedoms. Additionally, in the 1980s and 1990s Carlos Ripoll wrote a series of pamphlets published by the Cuban-American National Foundation on censorship and Stalinist methods in Cuba that, while perhaps biased against the Castro government, offer an interesting discussion and examples of limitations on artistic freedom. Roger Reed's *Cultural Revolution in Cuba* (1991) is a comprehensive study of how the Castro government has influenced cultural production.

10. Anthony Kerrigan's "What Are the Newly Literate Reading in Cuba? An 'Individualist' Memoir" speaks directly to the question of what was available to the people to be read in Cuba. Kerrigan published this essay in 1989 and it reflects his experience looking for certain books in Havana in the summer of 1986. Though this is mostly an anecdotal essay without a long term study that upholds his informal findings, it is an interesting account of what could be found easily in Havana in the middle of the 1980s, a time of a bit more economic and political stability than that of the 1970s, though it was also a time on the edge of great change that would result from the end of the Soviet Union and the Eastern bloc at the end of the 1980s and the beginning of the 1990s.

11. This speech took place in June of 1961 at the National Library in Havana. For the complete text, see *http://www.min.cult.cu/historia/palabrasalosintelectuales.html*. Hugh Thomas has a study of the importance of this speech for Cuba and Cuban intellectuals in *Cuba, or, The Pursuit of Freedom* (1465–1473). Chapter five of Marifeli Pérez-Stable's book analyzes the political context in the decade of the 1960s and Roberto González Echevarría's essay considers literature and its criticism in Revolutionary Cuba (Pérez-Stable 98–120; González Echevarría). Roberto González Echevarría, "Criticism and Literature in Revolutionary Cuba." *Cuba, Twenty-five Years of Revolution, 1959–1984*, ed. Sandor Halebsky and John M. Kirk (New York: Praeger Publishers, 1985) 154–173. Marifeli Pérez-Stable, *The Cuban Revolution* (New York: Oxford University Press, 1993). Hugh Thomas, *Cuba, or, The Pursuit of Freedom* (New York: Da Capo Press, 1998).

12. Hugh Thomas' *Cuba, or The Pursuit of Freedom* is the seminal work on Cuba from the eighteenth century to the first decades of the Revolution of 1959. His chapter "The Guardians" examines how the Revolution constructed itself in the 1960s. Thomas 1463

13. As quoted in Matas 434–435.

14. It is interesting to note that in the 1930s the Soviet censors more and more turned their attention to eradicating ambiguity in the texts. Jan Plamper, in the article "Abolishing Ambiguity: Soviet Censorship Practices in the 1930s," explores how one can see "an obsession with reducing signs to a single meaning"

(526). Similarly, in the latter half of the 1960s, the Cuban government began to refine more and more what it meant to write within the Revolution. We can see this in how the literary contests' rules defined unequivocally what made a revolutionary text. For example, despite its earlier award-winning appearances, absurdist theater would be no longer acceptable, seen as counterrevolutionary (Virgilio Piñera's *Dos viejos pánicos* being the most well-known having won the Casa de las Américas prize for theater in 1968). Nevertheless, when making these differences, it is important to remember the differences in sheer numbers between Soviet and Cuban censorship. The amount of victims in the Soviet example number in the millions; this is not the case for Cuba given the size of the country. The connections between Cuban and Soviet practices have been the question of many inquiries, but the early 1970s offer the most evidence of an association between the two. Carlos Ripoll in his *Los fundamentos del estalinismo en Cuba*, concludes that Cuban communism has strongly been inspired by the methods of Joseph Stalin and that this doctrine was prolonged only through tyranny.

15. Cabrera Infante had managed to publish in Spain a version of his *Tres tristes tigres* in 1967, for which he won the Premio Biblioteca Breve in Spain. However, the portrayal of sex in the novel is considered by some to have led to his loss of work with the Cuban government and his defection in 1965.
16. Guillermo Cabrera Infante, *Mea Cuba*, trans. Kenneth Hall (New York: Farrar Straus Giroux, 1994) 68–69.
17. The *Zafra de los diez millones* was an effort by the Cuban government to harvest ten million tons of sugar in 1970, what would be the biggest harvest ever. This was framed as an undertaking which would provide a moral victory at a time when the Revolution was being questioned at home by figures such as Heberto Padilla and others. However, the harvest, while still the largest ever, did not reach ten million tons and was thus seen as a failure.
18. It must also be pointed out that Cuba is still exercising under the same government of the 1960s and, therefore, access to papers and documents about the process of censorship is not available. It is still to be seen how much Fidel Castro's resignation of February 19, 2008 will affect access to archival records of the 1960s and 1970s.
19. In her essay "Literature and Society," Lourdes Casal points out that while royalties did not previously represent a large income for Cuban artists, receiving a salary from the state increased their dependence on the government (457). This can, obviously, be seen to curtail artistic freedoms.
20. The censorship and repression that surrounded much of the literary production of the late 1960s and especially that of Arenas is detailed in "Gays and the Cuban Revolution: The Case of Reinaldo Arenas" by Rafael Ocasio. It is interesting to note, however, that it is not only in the United States that there has been a renewed interest in these years. Raúl Castro's government has allowed more self-examination, including the Casa de las Américas conference "The Cultural Politics of the Revolutionary Period: Memory and Reflection," of which Ambrosio Fornet's discussion of the *quinquenio gris* formed a part.

21. Casal 457.
22. For Dopico Black, José Lezama Lima represents the third category of censorship: "marginally tolerated but subtly controlled" works (128). Because of his international renown, Lezama Lima's works could not be simply suppressed or censored as would happen to Reinaldo Arenas.
23. It appears that there was a reading of *Dos viejos pánicos* in Cuba when it was awarded the Casa de las Américas prize for theater (Forster 105). This will be explored in more detail at the end of this chapter.
24. Virgilio Piñera, *Teatro completo*, ed. Rine Leal (La Habana: Editorial Letras Cubanas, 2002) Act 1: 479. Text references are to the act and the page of this edition. Translations are mine. Hereafter, the quotes will be cited parenthetically with reference to the act and the page where they can be found.
25. The most informative articles on these two points come from Dara Goldman and José Quiroga. Dara Goldman, "Los límites de la carne: Los cuerpos asediados de Virgilio Piñera," *Revista Iberoamericana* 69 (2003): 1001–1015. José Quiroga, "Fleshing Out Virgilio Piñera from the Cuban Closet," *¿Entiendes? Queer Readings, Hispanic Writings*, ed. Emilie L. Bergmann and Paul Julian Smith (Durham, NC: Duke University Press, 1995). José Quiroga, "Virgilio Piñera: On the Weight of the Insular Flesh," *Hispanisms and Homosexualities*, ed. Silvia Molloy (Durham, NC: Duke University Press, 1998).
26. José Quiroga's essay "Piñera inconcluso," included in Rita Molinero's collection on Piñera, provides a particularly useful account of the re-introduction of Piñera's work after his death. Rita Molinero, ed., *Virgilio Piñera: La memoria del cuerpo* (San Juan, PR: Editorial Plaza Mayor, 2002).
27. Ana García Chichester, "Virgilio Piñera and the Formulation of a National Literature," *CR: The New Centennial Review* 2.2 (2002): 235. Emphasis from original.
28. García Chichester 236.
29. Leal vi.
30. Raquel Carrió Mendía, "Estudio en blanco y negro: Teatro de Virgilio Piñera" *Revista Iberoamericana* 56 (1990): 878.
31. Antonin Artaud, *El teatro y su doble*, trans. Enrique Alonso and Francisco Abelenda (Barcelona: Edhasa, 1978) and *Theater and Its Double*, trans. Mary Carolina Richards (New Cork: Grove Press, 1958). Martin Esslin, *The Theatre of the Absurd* (Woodstock, NY: Overlook Press, 1973).
32. Artaud, "No More Masterpieces," *Theater and Its Double* 79 and Esslin "Violence in Modern Drama" 177.
33. Zalacaín 15–60; Aguilú 35–62. Raquel Aguilú de Murphy, *Los textos dramáticos de Virgilio Piñera y el teatro del absurdo* (Madrid: Editorial Pliegos, 1989). Daniel Zalacaín, *El teatro absurdista hispanoamericano* (Valencia, España: Ediciones Albatros Hispanófila, 1985).
34. Woodyard; Palls; Martin. Eleanor Jean Martin, "*Dos viejos pánicos*: A Political Interpretation of the Cuban Theater of the Absurd," *Revista/Review Iberoamericana* 9 (1979): 50–56. Terry L. Palls, "El teatro del absurdo en Cuba: El compromiso artístico frente al compromiso político," *Latin*

American Theatre Review 11.2 (1978): 25–32. George Woodyard, "The Theatre of the Absurd in Spanish America," *Comparative Drama* 111.3 (1969): 183–192.
35. Woodyard 186.
36. Palls 25.
37. Palls 30.
38. Martin 55.
39. Martin 56.
40. Diana Taylor, *Theatre of Crisis: Drama and Politics in Latin America* (Kentucky: University Press of Kentucky, 1991) 9.
41. Taylor 9.
42. As quoted in Matías Montes Huidobro's study of Cuban theater. Montes Huidobro, Matías, *Persona, vida y máscara en el teatro cubano* (Miami: Ediciones Universales, 1973) 212.
43. Piñera's theater often plays with names to make a deeper suggestion about the characters and the play's arguments. This employment can also be seen in José Triana's landmark *La noche de los asesinos* (1965), where the grown siblings are named Cuca, Beba, and Lalo—an act that emphasizes their status as children and their desire to emancipate themselves from their parents. Triana, José, *La noche de los asesinos* (Madrid: Cátedra, 2001).
44. The *planilla*, while difficult to translate, refers to a sort of questionnaire that was common in Cuba in the 1960s. As in the play, a *planilla* contained a number of official questions that were to be answered and then handed back to the official department from which it was issued.
45. Montes Huidobro (1973) 212–213.
46. José Corrales, "Los acosados, Tabo, Tota, Montes Huidobro y Piñera." *Círculo: Revista de cultura* 29 (2000): 122. Translation mine.
47. "For example, I can read 'dos viejos pánicos' as 'two old people, filled with fear and apprehension,' and thus put an initial emphasis on the human figures. At the same time, one might read it as 'two old fears,' and in that way put stress on fear as an ancient emotional state related to the mythic god Pan" (Forster 106).
48. *Diccionario de la lengua española*, ed. La Real Academia Española <http://www.rae.es/>. Translation mine.
49. Corrales 120. Similarly, Forster's article details the connection between Arrabal and Piñera in a footnote. Merlin H. Forster, "Games and Endgames in Virgilio Piñera's *Dos viejos pánicos*," *In Retrospect: Essays on Latin American Literature*, ed. Elizabeth S. Rogers and Timothy J. Rogers (South Carolina: Special Literary Publications, 1987) 113–114.
50. As quoted in Francisco Torres Monreal, "Introducción," *Teatro pánico*, By Fernando Arrabal (Madrid: Cátedra, 1986) 15. Emphasis from the original.
51. Torres Monreal 24.
52. Torres Monreal 19.
53. Torres Monreal 38.
54. Cabrera Infante 68–69.

55. Artaud 82.
56. Severino João Albuquerque, *Violent Acts: A Study of Contemporary Latin American Theatre* (Detroit: Wayne State University Press, 1991) 83.
57. Torres Monreal 43.
58. Forster 107. In his essay, Forster makes these references in order to connect Piñera's play with Samuel Beckett's *Endgame* and suggest the former play as "a significant example of accommodation and transformation" of theater of the absurd (105).
59. While Torres Monreal points out that circularity and repetition are characteristics of much absurd and modern literature, he establishes two types of circularity in Arrabal: "En Arrabal distinguiríamos dos tipos de obras: las *unicirculares* (el final vuelve al principio) y las *pluricirculares* (en el interior de la obra, la misma estructura vuelve sobre sí mismo en repetidas ocasiones) [In Arrabal we see two types of plays: the *unicirculars* (the end returns to the beginning) and the *pluricirculars* (in the interior of the play, the same structure returns to itself repeatedly]" (46).
60. See Katherine Ford, "El espectáculo revolucionario: El teatro cubano de la década de los sesenta," *Latin American Theatre Review* 39 (2005): 95–114. In this essay, the word "revolution" is used in the double sense of a break with the past and as a repetition to point out the duality that Cuban theater is seeing in the context of the 1960s.
61. Ford 97.
62. Ford 105–106.
63. Woodyard 188.
64. Montes Huidobro (1973) 435.
65. Torres Monreal 52.
66. Torres Monreal 52.
67. Torres Monreal 53.
68. This attempt to break with the past through "death" is reminiscent of the siblings' attempt to break with their tyrannical parents through "murder" in *La noche de los asesinos*. Here, the three siblings rehearse their parents' murder and imagine their own reactions to this horrific act. Inherent in these imaginings exists a desire to begin again, though this is truncated by the fact that these actions are never fulfilled but remain imagined.
69. The use of light here remembers the use of light in Piñera's "La isla en peso," the essential poem that considers the effect of the island's insularity on constructing the poet. In the poem, the light, seen in the dazzling sunlight of the day, like the water that surrounds the island, weighs heavy and overwhelming on the bodies that it takes in. Matías Montes Huidobro discusses the importance of light in Piñera's work. He believes that it is at this moment that fear becomes another personality within the play and connects this with Piñera's *Electra Garrigó*. Piñera, Vigilio, "La isla en peso," *La vida entera* (La Habana: UNEAC, 1968) 25–42. Montes-Huidobro (1973) 436–437.
70. Girard 10.

2 COBWEBS OF MEMORY: HISTORY MADE WITH VIOLENCE IN ABELARDO ESTORINO'S *LA DOLOROSA HISTORIA DEL AMOR SECRETO DE DON JOSÉ JACINTO MILANÉS* (1974)

1. This is, of course, a simplification of the complex and multi-faceted circumstances in which Cuban independence was advocated. Hugh Thomas's essay "La colonia española de Cuba" in *Historia del Caribe* offers a much more detailed and richly-hued conversation on Cuba's continued status as a colony. Hugh Thomas, "Capítulo 2: La colonia española de Cuba," In *Historia del Caribe* (Barcelona: Editorial Crítica, 2001) 39–55.
2. Thomas, *Historia del Caribe* 47; Tulio Halperín Donghi, *The Contemporary History of Latin America*, ed. and trans. John Charles Chasteen (Durham, NC: Duke University Press, 1993) 156.
3. Hugh Thomas, *Cuba, or The Pursuit of Freedom* (New York: Da Capo Press, 1998) 205–206. For a more detailed discussion of the Conspiracy and its implications both in the nineteenth century and beyond, see Robert L. Paquette's *Sugar Is Made with Blood: The Conspiracy of la Escalera and the Conflict between Empires over Slavery in Cuba* (Middletown, CT: Wesleyan University Press, 1988).
4. A record of what happened at this Congress can be found in the article "Primer congreso nacional de educación y cultura" in the *Gaceta de Cuba*. "Primer congreso nacional de educación y cultura." *Gaceta de Cuba* 90–91 (March–April 1971): 2–13.
5. Paul Christopher Smith, "Theatre and Political Criteria in Cuba: Casa de las Américas Awards, 1960–1983," *Cuban Studies/Estudios cubanos* 14.1 (1984): 43–47.
6. Lourdes Casal maintains that Leopoldo Ávila was most likely José Antonio Portuondo, a well-known Revolutionary critic who would also preside over Padilla's self-criticism at UNEAC headquarters. However, Roger Reed and Fornet himself assert that this was really Luis Pavón, the magazine's director. It seems likely that Fornet, having written most recently among other factors, would have the most access to up-to-date information. Lourdes Casal, "Literature and Society," *Revolutionary Change in Cuba*, ed. Carmelo Mesa-Lago (Pittsburgh: University of Pittsburgh Press, 1971) 20; Roger Reed, *The Cultural Revolution in Cuba* (Geneva: Latin American Round Table, 1991) 106; Ambrosio Fornet, "El Quinquenio gris: Revisitando el término" (Havana 2006. http://www.criterios.es/pdf/fornetquinqueniogris.pdf) 9.
7. For more information on how Estorino's plays fit into realist theater, see George Woodyard's essay on Estorino's theater. In this article, Woodyard examines how Estorino's plays fit into a realist vein as proposed by Arnold Kettle's article of Charles Dickens' *Bleak House*. George Woodyard, "Estorino's Theatre: Customs and Conscience in Cuba," *Latin American Literary Review* 11.22 (1983): 57–63.

8. The information about Milanés supplied here comes from Salvador Arias' *Tres poetas en la mirilla*, *Historia de la literatura cubana: Tomo 1*, and Max Henríquez Ureña's *Panorama histórico de la literatura cubana*. Salvador Arias, *Tres poetas en la mirilla* (La Habana: Editorial Letras Cubanas, 1981); Max Henríquez Ureña, *Panorama histórico de la literatura cubana* (La Habana: Editorial Arte y Literatura, 1978); *Historia de la literatura cubana: Tomo 1: La colonia: desde los orígenes hasta 1898* (La Habana: Editorial Letras Cubanas, 2002).

9. Domingo del Monte y Aponte (1804–1853) was born in Venezuela, son of Dominicans, and moved to Cuba with his family in 1810. After having traveled around Europe, del Monte returned to Cuba in 1829 and, in the following decade, became an important influence on Cuban intellectuals and Romanticism for his writings, but perhaps even more for his personal exchanges. He was a strong animator of his fellow intellectuals in both public and private meetings, a role that can be observed in the scene "Tertulia" in *La dolorosa historia*. Like many other Cuban intellectuals from this historical moment, he ceaselessly debated independence, though his perspective on it was much more complicated than that of others, given his own familial ties to the Cuban oligarchy that wanted to maintain the status quo and his loyalty to Spanish culture and tradition. On the issue of abolition, del Monte's ideas were much clearer and he used his famous *tertulias* to urge other writers and intellectuals to protest against slavery. In fact, del Monte's *tertulia* included Juan Francisco Manzano, an ex-slave who was freed thanks to a collection of funds by the other *tertulia* members. Manzano's *Autobiografía* and some poems were later published in London, sent there with other writings against the slave trade. Del Monte's writings and influence caused difficulty for him and in 1843 he left Cuba with his family. When the *Conspiración de la Escalera* broke out, other intellectuals opposing the slave trade returned to Cuba to defend themselves, but del Monte did not, a decision that is argued over with and repudiated by Milanés in the play. Del Monte died in Madrid in 1853 (Henríquez Ureña 190–191; Arias [*Historia de la literatura cubana: Tomo1*] 141–151)

10. Milanés' play is about the Conde Alarcos, who, though Spanish, had pledged his allegiance to the French king, and had consequently been betrothed to his daughter, Blanca. Before Alarcos departs on a pilgrimage to Santiago de Compostela, Blanca gives her honor to the count, thus unofficially marrying the two. While in Spain, Alarcos marries and has children with a woman from Seville, Leonor. Nevertheless, he must keep his word to the French king to return to Paris, where it is found out that he has married another woman, being all but officially married already to Blanca. The French king orders him to kill his Spanish wife in order to marry Blanca and, though he detests the chains that bind him to the king and tries to save Leonor, she is killed by an executioner, proving that they are slaves subject to the king's desires. José Jacinto Milanés, *El conde Alarcos*, In *Teatro del siglo XIX* (La Habana: Editorial Letras Cubanas, 1986).

11. Antón Arrufat, *Virgilio Piñera: entre él y yo* (La Habana: UNEAC, 1994) 41–47.

12. Vivian Martínez Tabares, "La dolorosa búsqueda de los recuerdos," *Teatro cubano contemporáneo* (Madrid: Sociedad Estatal Quinto Centenario, 1992) 349.
13. Jorge Febles, "Recontextualización poemática en La dolorosa historia del amor secreto de don José Jacinto Milanés," *Latin American Theatre Review* 31.2 (1998), 79.
14. Febles 81.
15. This is taken from an interview with Abelardo Estorino in May of 2007. When asked about the connection between writing and premiering, he answered that he had had much success in the Havana theaters publishing what he wrote until *La dolorosa historia*: "Todas mis obras se han estrenado durante su tiempo por otros directores hasta que yo tuve un fracaso con mi primera obra de Milanés. Porque tú sabes que yo tengo dos. Y mi obra primera la tomó un director muy importante cubano que es Vicente Revuelta. Él hizo un montaje muy experimental, en un espacio que no era un espacio convencional del teatro italiano sino lo ensayó en un patio. La ensayó así porque el teatro que teníamos, que es el Hubert de Blanck que era donde trabajábamos, estaba en reparación. Cuando las reparaciones terminaron, era el turno para que él la llevara para el teatro pero él tardó mucho en convertir esa obra que estaba montada en un espacio, convertirla en otra cosa. Mientras tanto, el teatro necesitaba abrir las puertas, ya estaba arreglado, y empezar a trabajar. Y él nunca llegó a estrenarla. Entonces, eso es lo que me dio a mí un pie para yo tratar de dirigir mis propias obras porque no quería que me pasara. [All of my plays have premiered in their time by other directors until I had a failure with my first play on Milanés (*La dolorosa historia*). Because you know I have two. The first was taken on by an important director, Vicente Revuelta. He did a very experimental staging, in a space that wasn't a conventional Italian theater space, but a patio. He rehearsed it there because the theater that we had, the Hubert be Blanck, was being repaired. When the reparations were done, it was his turn to put on the play, but he took so long to change it to this space. In the meantime, the theater needed to open its doors and start work. So he never premiered it. That's when I decided to direct my own works because I didn't want that to happen.]"
16. Estorino returned to the figure of José Jacinto Milanés in 1993 with *Vagos rumores*, a play that Estorino himself sees as a re-write and clarification of *La dolorosa historia*, as he told me in an interview conducted on May 8, 2007. *Vagos rumores* tells the same story of Milanés, but with less detail and characters. In this second play, Estorino has three characters: Milanés, the Mendigo, and Carlota, Milanés' sister. The Mendigo and Carlota take on other characters within the play where, in *La dolorosa historia*, Estorino would have used another actor. The argument of this second play is much the same as the first, though it premiered soon after it was written in the Hubert de Blanck theater.
17. Abelardo Estorino, *Memorias de Milanés* (Matanzas, Cuba: Ediciones Matanzas, 2005) Prólogo 28. This collection was a commerative edition that published *La dolorosa historia del amor secreto de José Jacinto Milanés* together

with the play *Vagos rumores*, which was another exploration of Milanés, though shorter. All text references will come from this edition. References are to the scene (designated with a one-word title) and the page of this edition. Translations are mine. Hereafter, the quotes will be cited parenthetically with reference to the scene and the page where they can be found.
18. Abelardo Estorino, personal interview. May 8, 2007.
19. Born Gabriel de la Concepción Valdés (1809–1844), Plácido was the illegitimate son of a Spanish dancer and a mulatto hairdresser, born in Havana. At his birth, he was abandoned by his mother, the dancer, and given to the care of his paternal grandmother. Not having enough money to study, he became an apprentice to a carpenter at age twelve, around the same time he began writing verses. He later practiced as a comb salesman in both Matanzas and Havana. He gained renown in 1834 with his poem "La siempreviva" and began to sign his poems with the pseudonym Plácido. José María Heredia invited him to move to Mexico in order to escape the limitations that the Cuban society placed on him because of his race, though he declined.
20. Matías Montes-Huidobro, "El discurso teatral histórico-poético de Abelardo Estorino: entre el compromiso y la subversión," *Alba de América* 9.16–17 (1991): 250.
21. Throughout *La dolorosa historia*, José Jacinto Milanés obsesses over the idea of being forgotten and this will be explored later on in this chapter. Nevertheless, this is an important connection to point out at this time, given its role throughout the play.
22. Montes-Huidobro, "El discurso teatral histórico-poético de Abelardo Estorino: entre el compromiso y la subversión," 250–251.
23. This is the year that many literary critics have identified a marked change in Milanés' poetry towards socio-political topics due, in part, to his relationship and correspondence with Domingo del Monte. Many critics have also noted that these poems tend to be more valuable for their intention than for their own achievements as poems, though "El mendigo" has been seen as an exception (Arias, *Tres poetas en la mirilla* 92).
24. All the quotes from "El mendigo" are from Salvador Arias' collection of poetry from the colonial period, for which I give the page numbers here. Translations are mine. Salvador Arias, ed., *Poesía cubana de la colonia* (La Habana: Editorial Letras Cubanas, 2002) 77–78.
25. Montes-Huidobro, "El discurso teatral histórico-poético de Abelardo Estorino: entre el compromiso y la subversión," 252.
26. Zequeira is, of course, Manuel de Zequeira y Arango (1764–1846), the Cuban poet and soldier. He was sent to South America in 1810 and fought in the wars of independence, defending the Spanish crown. Throughout all of this, he continued to write poetry until his own loss of reason in 1821, a state in which he remained until his death. With Zequeira, along with the poets Manuel Justo Rubalcava and Manuel María Pérez y Ramírez, a new period of Cuban poetry began, marked by a personal tone connected with the social reality of the moment. Historically, his poetry is associated with the neoclassical period.

"A la piña" is his most anthologized poem and is one of the early songs in celebration of the Cuban flora (Henríquez Ureña 99–103; *Historia de la literatura cubana: Tomo 1* 70–80).
27. The first four lines of the scene quoted are analyzed in Matías Montes-Huidobro's essay as evidence of the merging between time periods, given that the question is asked in the present ("llega [arrives]"), referring to the smell of flowers in the past ("había [there were]"). This is an excellent example that solidifies Montes-Huidobro's argument about the anti-temporality in the play. Nevertheless, it is not my purpose here to discuss temporality but the use of poetry in the dramatic work.
28. Montes-Huidobro, "El discurso teatral histórico-poético de Abelardo Estorino: entre el compromiso y la subversión," 247.
29. Montes-Huidobro, "El discurso teatral histórico-poético de Abelardo Estorino: entre el compromiso y la subversión," 254.
30. Federico Milanés did, in fact, dedicate himself to preserving his brother's literary work and memory. He was a playwright and poet in his own right. Nevertheless, none of his plays has survived and his most-anthologized poem is "Aniversario," written in memoriam of his late brother.
31. Del Monte, as stated before, refers to the nineteenth-century intellectual Domingo del Monte, the man who held the *tertulias*. Cirilo Villaverde (1812–1894) is the author of the influential Romantic novel *Cecilia Valdés*. Ramón de Palma y Romay (1812–1860) is another important Romantic novelist who wrote "Matanzas y Yumurí," which is said to be the first indigenous narrative in Cuba. He is considered, together with Villaverde, as the initiator of the narrative form in Cuba (Henríquez Ureña 280–286, 288–289).
32. Here the literary men mainly discuss a written theater that would premiere in smaller or less accessible theaters. To understand the workings of the popular theater that was largely seen and available in Cuba during the nineteenth century, see Jill Lane's *Blackface Cuba*. In this excellent examination of theater in nineteenthth century Cuba, Lane details and examines the connection between blackface performance, a budding sense of nationalism and populist theater. Jill Lane, *Blackface Cuba, 1840–1895* (Philadelphia: University of Pennsylvania Press, 2005).
33. It must be remembered that because of his health Milanés was not able to travel to Havana for the premiere of his play given his mental condition.
34. Lane 52.
35. In *Torture and Truth*, Page duBois' states that the Greek word for torture "means first of all a touchstone used to test gold for purity; the Greeks extended its meaning to denote a test or trial to determine whether something or someone is real or genuine (7)" This suggests that torture was then used as a way to test if a statement was true, meaning that a slave's testimony would only be accepted if he or she had been tortured. Page duBois, *Torture and Truth* (New York: Routledge, 1991) 1–8.
36. duBois 47.
37. Herbert Lindenberger, *Historical Drama: The Relation of Literature and Reality* (Chicago: University of Chicago Press, 1975) 45–46.

3 Filming the Bourgeoisie: Defining Identity with Violence in Eduardo Pavlovsky's *La mueca* (1970)

1. Matías Montes-Huidobro, "Psicoanálisis fílmico-dramático de *La mueca*," *Monographic Review/Revista Monográfica* 7 (1991): 298–299.
2. Jorge Dubatti, "Estudio preliminar," *Teatro completo I*, de Eduardo Pavlovsky, Buenos Aires: Atuel, 1997. For a more detailed account of the expressions of Argentina's Theater of the Absurd, see Angela Blanco Amores de Pagella "Manifestaciones del teatro del absurdo en Argentina," *Latin American Theatre Review* 8.1 (1974): 21–24. All translations are mine.
3. Eduardo Pavlovsky, "Algunos conceptos sobre el teatro de vanguardia," *Teatro del '60* (Buenos Aires: Ediciones Letra Buena, 1992) 114–115. Translations mine.
4. Pavlovsky, "Algunos conceptos sobre el teatro de vanguardia" 115. I purposely use "man" here since that is what Pavlovsky himself uses.
5. Dubatti sees that the 1960s need to be divided into two separate moments with the Mayo Francés in 1968 separating the first from the second (Jorge Dubatti, "Estudio preliminar," *Teatro completo II*, de Eduardo Pavlovsky [Buenos Aires: Atuel, 1998] 10–11).
6. Dubatti (1997) 14; Dubatti (1998) 11.
7. In 1973, and then again in 1983 and 1989, he presented himself as a candidate for, first, *Partido Socialista de los Trabajadores* (PST) and, then, *Movimiento al Socialismo* (MAS).
8. Arancibia, Juana A and Zulema Mirkin, *Volumen II: Teatro argentino durante el proceso (1976–1983): ensayos críticos-entrevistas* (Buenos Aires: Instituto Literario y Cultural Hispánico, 1992) 223.
9. Teatro Abierto was a theatrical response to the Proceso de Reorganización Nacional when the governement repressed any cultural production that went against their national product. Sectors like radio, television, cinema and theater were severely restricted to oppressive norms in order to maintain them within the government's ideology (Arancibia et al. 16–18). In 1981, in a desire to eliminate the isolation in which Argentine theater had fallen, the government opened la Comedia Rosarina in Rosario without the intervention or help of any of those most closely involved with theater in the past. As an answer to this, Osvaldo Dragún and others involved in the theater community initiated Teatro Abierto, a movement that had three fundamental goals: "1) Demostrar la existencia y la vitalidad de un teatro muchas veces negadas. 2) Recuperar un público que se había perdido para el teatro en general, a través de una programación con gran variedad de estilos y libertad temática. 3) Cobrar precios bajísimos" [1)Demonstrate the existence and vitality of a theater often denied. 2) Recuperate an audience that had been lost to the theater in general, through programming that is varied in style and thematic liberty. 3) Charge very inexpensive prices] (Arancibia et al. 21). Teatro

Abierto was a phenomenon repeated in the following years and helped to begin to recuperate voices that had been violently pushed to the margins.
10. Alfonso De Toro, "El teatro *menor* postmoderno de Eduardo Pavlovsky o el 'Borges/Bacon' del teatro: De la periferia al centro," *Gestos* 31 (Abril 2001): 101. Although this quote examines Pavlovsky's theater more generally, de Toro's article admirably studies how this collective process unfolded for the specific example of Pavlovsky's *Poroto*.
11. David Rock, *Argentina 1516–1982: From Spanish Colonization to the Falklands War* (Los Angeles: University of California Press, 1985) 359.
12. Rock 360.
13. Eduardo Pavlovsky, *La mueca* (Buenos Aires: Ediciones Búsqueda, 1988) Act 1: 16, 18.
14. Pavlovsky *La mueca* Act 2: 37. Text references are to the act and the page of this edition. Translations are mine. Hereafter, the quotes will be cited parenthetically with reference to the act and the page where they can be found.
15. *Diccionario de la lengua española*, ed. La Real Academia Española <http://www.rae.es/>.
16. Charles B. Driskell, "Power, Myths and Aggression in Eduardo Pavlovsky's Theater," *Hispania* 65.4 (1982): 573.
17. Richard Schechner, *Performance Theory* (New York: Routledge, 2003) 191.
18. George O. Schanzer, "El teatro vanguardista de Eduardo Pavlovsky," *Latin American Theatre Review* 13.1 (1979): 10.
19. Schanzer 10.
20. Schanzer 10.
21. Montes-Huidobro, "Psicoanálisis fílmico-dramático de *La mueca*," 298–299.
22. Laura Mulvey, "Visual Pleasure and Narrative Cinema," *Visual and Other Pleasures* (Bloomington, IN: Indiana University Press, 1989) 16–17.
23. Mulvey 17.
24. Mulvey 17.
25. Mulvey 17.
26. Mulvey 19.
27. De Toro 102.
28. Martin Esslin, "Violence in Modern Drama," *The Theatre of the Absurd* (Woodstock, NY: Overlook Press, 1973) 177–178.
29. Marvin Carlson, in his essay "What Is Performance?" for *The Performance Studies Reader*, finds three separate definitions of performance: "So we have two rather different concepts of performance, one involving the display of skills, the other also involving display, but less of particular skills than of a recognized and culturally coded pattern of behavior. A third cluster of usages takes us in rather a different direction. When we speak of someone's sexual performance or linguistic performance or when we ask how well a child is performing in school, the emphasis is not so much on display of skill (although that may be involved) or on the carrying out of a particular pattern of behavior, but rather on the general success

of the activity in light of some standard of achievement that may not itself be precisely articulated (70)." For this play, the word 'performance' is being used in conjunction with the second definition: the idea of a pattern of behavior that is both recognized and coded socially. Marvin Carlson, "What Is Performance?" *The Performance Studies Reader* (New York: Routledge, 2004).
30. Dubatti (1997) 14.
31. This questioning of the accepted manner of events can also be seen in José Triana's *La noche de los asesinos*, the Cuban play in which Lalo and Cuca clash over their different ways of thinking:

> CUCA: El cenicero debe estar en la mesa y no en la silla.
> LALO: Haz lo que te digo.
> CUCA: No empieces, Lalo.
> LALO: (*Coge el cenicero y lo coloca en la silla*). Yo sé lo que hago. (*Apuña el florero y lo instala en el suelo*). En esta casa el cenicero debe estar encima de una silla y el florero en el suelo. (76)

32. Mikhail Bakhtin, "Epic and Novel," *The Dialogic Imagination*, ed. Michael Holquist, trans. Caryl Emerson and Michael Holquist (Austin: University of Texas Press, 1981) 23.
33. Gary Saul Morson and Caryl Emerson, *Mikhail Bakhtin: Creation of a Prosaics* (Stanford: Stanford University Press, 1990) 443.
34. De Toro 102.
35. Mikhail Bakhtin, *Rabelais and His World*, trans. Hélène Iswolsky, (Bloomington: Indiana University Press, 1984) 7.
36. Claudia Gilman, *Entre la pluma y el fusil: debates y dilemmas del escritor revolucionario en América Latina* (Buenos Aires: Siglo veintiuno editores, 2003) 58.
37. Gilman 51.
38. Rock 355.
39. Gilman 66.
40. This connection between class and authority recalls the famous Argentine short story "El matadero" from Esteban Echevarría in that the intruder's obvious social status does not afford him a position of authority in the neighborhood of Buenos Aires that he has stumbled upon. Instead, the greater number of those who "belong" in that area allows them to exercise authority over the intruder. Here, as in *La mueca*, numbers dictate authority, not social class, as Helena would like to believe. Esteban Echeverría, "El matadero," *Antología del cuento hispanoamericano*, ed. Fernando Burgos (México: Editorial Porrúa, 1991) 1–16.
41. Judith Butler, "Performative Acts and Gender Constitution: An Essay in Phenomenology and Feminist Theory," *Performing Feminisms: Feminist Critical Theory and Theatre*, ed. Sue Ellen Case (Baltimore: Johns Hopkins University Press, 1990) 278.
42. Judith Butler, *Excitable Speech: A Politics of the Performative* (New York: Routledge, 1997) 2.

4 Disorderly Conduct: The Violence of Spectatorship in Griselda Gambaro's *Información para Extranjeros* (1973)

1. Susana Tarantuviez, *La escena del poder*. *El teatro de Griselda Gambaro*, Buenos Aires: Corregidor, 2007 (332).
2. Griselda Gambaro's theater is the subject of hundreds of articles and numerous books. Diana Taylor is one of the most well-known scholars of Gambaro's drama, though the Marguerite Feitlowitz's Introduction to the English translation of three of her plays stands out as a study. Also, the recent publication of Susana Tarantuviez's work on Gambaro's theater is an important addition to this growing body of scholarship. See Diana Taylor, *Disappearing Acts: Spectacles of Gender and Nationalism in Argentina's "Dirty War,"* (Durham, NC: Duke University Press, 1997), her essay "Paradigmas de crisis: La obra dramatic de Griselda Gambaro" in the collection *En busca de una imagen: Ensayos críticos sobre Griselda Gambaro y José Triana*, ed. Diana Taylor, Canadá: Girol Books, 1989) and *Theatre of Crisis: Drama and Politics in Latin America* (Kentucky: University Press of Kentucky, 1991). Marguertite Feitlowitz, "Crisis, Terror, Disappearance: The Theater of Griselda Gambaro," In *Information for Foreigners* (Illiniois: Northwestern University Press, 1992). Susana Tarantuviez, *La escena del poder*. *El teatro de Griselda Gambaro* (Buenos Aires: Corregidor, 2007).
3. Despite its difficult history of representation, *Información* has been the subject of much research and discussion in academic circles, especially in the US. Some names that stand out as most important in reference to this play are Jason Cortés, Myriam Yvonne Jehenson, Mady Schutzman, John Fleming, Rosalea Postma, and Dick Gerdes. Jason Cortés, "La teatralización de la violencia y la complicidad del espectáculo de *Información para extranjeros* de Griselda Gambaro," *Latin American Theatre Review* 35.1 (2001): 47–61. Myriam Yvonne Jehenson, "Staging Cultural Violence: Griselda Gambaro and Argentina's 'Dirty War'," *Mosaic: A Journal for the Interdisciplinary Study of Literature* 32.1 (1999): 85–104. Mady Schutzman, "Calculus, Clairvoyance, and Communitas," *Women and Performance: A Journal of Feminist Theory* 10 (1999): 117–33. John Fleming, "Argentina on Stage: Griselda Gambaro's *Information for Foreigners* and *Antígona furiosa*," *Selected Proceedings: Louisiana Conference on Hispanic Languages and Literatures*, ed. Joseph V. Ricapito (Baton Rouge: Louisiana State University Press, 1994). Rosalea Postma, "Space and Spectator in the Theatre of Griselda Gambaro: *Informacion para extranjeros*," *Latin American Theatre Review* 14.1 (1980): 35–45. Dick Gerdes, "Recent Argentine Vanguard Theatre: Gambaro's *Información para extranjeros*," *Latin American Theatre Review* 11.2 (1978): 11–16.
4. This alteration of the traditional, linear plot that moves from page one through to the end can also be seen in Julio Cortázar's *Rayuela*, where the reader is invited to jump from chapter to chapter in an apparent lack of order. Here,

however, Gambaro's written script has no *Tablero de dirección* to lay out her intended direction for reading for the reader. Instead, the reader, upon her own initiative is invited to move through the scenes and to stage a mental order according to her own ideas and considerations. The physical representation, in turn, does progress upon an order prescribed by the guide of the group of spectators, but this also differs from Cortázar's layout in *Rayuela*. Cortázar is the one who dictates the order in his monumental novel. Gambaro's play, like all theater, encourages a collaboration between all elements involved—the playwright, the actors, the director, the audience—that, no matter how small in one particular production, is greater than that of a novel. Julio Cortázar, *Rayuela* (Madrid: Cátedra, 2004).
5. Griselda Gambaro, *Teatro 2. Dar la vuelta. Información para extranjeros. Puesta en claro. Sucede lo que pasa* (Buenos Aires: Ediciones de la Flor, 1987) 68. Griselda Gambaro, *Information for Foreigners: Three Plays by Griselda Gambaro*, ed. and trans. Marguerite Feitlowitz (Illinois: Northwestern University Press, 1992) 67. All text references in Spanish will come from the former and in English are from the translation by Marguerite Feitlowitz. References are to the scene and the page. The scene is the same in both editions and is not repeated; the page numbers are different. Hereafter, the quotes will be cited parenthetically with reference to the scene and the pages where they can be found.
6. *Diccionario de la lengua española*, ed. La Real Academia Española <http://www.rae.es/>.
7. Rosalea Postma identifies how the role of space becomes a central issue in the play's objectives in her article: "The entire theatrical space becomes the staging area, and therefore the spectator is engulfed by the dramatic action. The structure of *Información* forces the spectator to struggle with the confinement of space. The spectator's traditionally passive role in 'going to the theatre' becomes active. While no audience reaction is fully predictable, every dramatic text functions to program a potential response; and while we have no performance data on *Información*, its script presumes performance. [...] The larger spaces in the set for *Información* are fragmented into smaller divisions, and the already narrow hallways are further restricted by the clutter of lockers of various sizes, each with a louvered door. Each guide maneuvers his group of spectators through these hallways, up and down stairways, in and out of rooms" (37).
8. Tarantuviez 201–202.
9. Antonin Artaud, *Theater and its Double*, trad. Mary Carolina Richards (New York: Grove Press, 1958) 96.
10. Gerdes 12.
11. Gerdes 12.
12. Postma, Rosalea 38.
13. Idelber Avelar, *The Letter of Violence: Essays on Narrative, Ethics, and Politics* (New York: Palgrave Macmillan, 2004) 3–4.
14. Postma 40.
15. Gambaro 70.

16. Taylor, *Disappearing Acts* 167.
17. Cortés 54. Translation mine.
18. Artaud 82.
19. Stanley Milgram, *Obedience to Authority: An Experimental View* (New York, HarperCollins Publishers, 1974) 3.
20. Milgram 3.
21. I use the masculine pronoun here to refer to the student because in the play it is "El joven."
22. Milgram 3–4.
23. Elaine Scarry, *The Body in Pain: The Making and Unmaking of the World* (New York: Oxford University Press, 1985) 4.
24. Cortés 50.
25. Diana Taylor's *Disappearing Acts* examines the use of performance and the manipulation of images during the Dirty War. This is an interesting discussion in light of *Información para extranjeros* considering that the play was written only three years before the Dirty War began.
26. Artaud 82.
27. Bertolt Brecht, *Brecht on Theatre: The Development of an Aesthetic*, ed and trans. John Willett (New York: Hill and Wang, 1964) 190.
28. Michel Foucault, *Discipline and Punish: The Birth of the Prison*, trans. Alan Sheridan (New York: Random House, 1977) 7–8.
29. Foucault 12.
30. Foucault 14–16.
31. Postma 41.

Bibliography

Adler, Heidrun y Adrián Herr, ed. *De las dos orillas: Teatro cubano*. Frankfurt: Vervuert, 1999.
Aguilú de Murphy, Raquel. *Los textos dramáticos de Virgilio Piñera y el teatro del absurdo*. Madrid: Editorial Pliegos, 1989.
Albuquerque, Severino João. *Violent Acts: A Study of Contemporary Latin American Theatre*. Detroit: Wayne State University Press, 1991.
Anderson, Thomas F. *Everything in Its Place: The Life and Works of Virgilio Piñera*. Lewisburg: Bucknell University Press, 2006.
Arancibia, Juana A and Zulema Mirkin. *Volumen II: Teatro argentino durante el proceso (1976–1983): ensayos críticos-entrevistas*. Buenos Aires: Instituto Literario y Cultural Hispánico, 1992.
Arendt, Hannah. *On Violence*. New York: Harcourt, Brace & World, 1970.
Arias, Salvador. *Tres poetas en la mirilla*. La Habana: Editorial Letras Cubanas, 1981.
Arias, Salvador, ed. *Poesía cubana de la colonia*. La Habana: Editorial Letras Cubanas, 2002. 77.
Aristotle. *Poetics*. Trans. Leon Golden. Tallahassee: Florida State University Press, 1981.
Arrufat, Antonin. *Los siete contra Tebas*. La Habana: UNEAC, 1968.
———. *Virgilio Piñera: entre él y yo*. Havana: UNEAC, 1994.
Artaud, Antonin. *El teatro y su doble*. Trans. Enrique Alonso y Franciso Abelenda. Barcelona: Edhasa, 1978.
———. *Theater and Its Double*. Trans. Mary Carolina Richards. New Cork: Grove Press, 1958.
Austin, J.L. *How to Do Things with Words*. Ed. J.O. Urmson and Marina Sbisà. Cambridge: Harvard University Press, 1975.
Ávila, Leopoldo. "Las provocaciones de Padilla." *Verde Olivo* 9:45 (1968) 17–18.
———. "Sobre algunas corrientes de la crítica y la literatura de Cuba." *La Gaceta de Cuba* Noviembre-Diciembre (1968) 3–4.
Avelar, Idelber. *The Letter of Violence: Essays on Narrative, Ethics, and Politics*. New York: Palgrave, 2004.
Bakhtin, Mikhail. "Epic and Novel." *The Dialogic Imagination*. Ed. Michael Holquist. Trans. Caryl Emerson and Michael Holquist. Austin: University of Texas Press, 1981.
———. *Rabelais and his World*. Trans by Hélène Iswolsky. Bloomington: Indiana University Press, 1984.

Benítez Rojas, Antonio. "Comments on Georgina Dopico Black's 'The Limits of Expression: Intellectual Freedom in Postrevolutionary Cuba'." *Cuban Studies/ Estudios cubanos* 20 (1990), 171–174.
Biron, Rebecca. *Murder and Masculinity: Violent Fictions of Twentieth Century Latin America.* Nashville: Vanderbilt University Press, 2000.
Blanco Amores de Pagella, Angela. "Manifestaciones del teatro del absurdo en Argentina." *Latin American Theatre Review* 8.1 (1974) 21–24.
Boal, Augusto. *Teatro del oprimido y otras poéticas políticas.* Buenos Aires: Ediciones de la Flor, 1974.
———. *Theatre of the Oppressed.* Trans. Charles A. & Maria-Odilia Leal McBride. New York: Theater Communications Group, 1985.
Brecht, Bertolt. *Brecht on Theatre: The Development of an Aesthetic.* Ed. John Willett. New York: Hill and Wang, 1992.
"Las Brigadas de Teatro de la Coodinación Provincial de Cultura de La Habana." *Conjunto* 2 (1964): 59–64.
Butler, Judith. *Bodies that Matter: On the Dirscursive Limits of "Sex."* New York: Routledge, 1993.
———. *Excitable Speech: A Politics of the Performative.* New York, Routledge: 1997.
———. "Performative Acts and Gender Constitution: An Essay in Phenomenology and Feminist Theory." *Performing Feminisms: Feminist Critical Theory and Theatre.* Ed. Sue Ellen Case. Baltimore: Johns Hopkins University Press, 1990.
Cabrera Infante, Guillermo. *Mea Cuba.* Barcelona: Plaza y Janes Editores, 1992.
El caimán barbudo 15 (1967).
Carlson, Marvin. "What is Performance?" *The Performance Studies Reader.* New York: Routledge, 2004.
Carrió Mendía, Raquel. "Estudio en blanco y negro: Teatro de Virgilio Piñera." *Revista iberoamericana* 56 (1990): 871–880.
Casal, Lourdes. "Literature and Society." *Revolutionary Change in Cuba.* Ed. Carmelo Mesa-Lago. Pittsburgh: University of Pittsburgh Press, 1971.
Case, Sue-Ellen, ed. *Performing Feminisms: Feminist Critical Theory and Theatre.* Baltimore: John Hopkins University Press: 1990.
Castañeda, Jorge G. *Utopia Unarmed: The Latin American Left After the Cold War.* New York: Alfred A. Knopf, 1993.
Castro, Fidel. "Discurso de clausura." *Casa de las Américas* 9.65–66 (March–June 1971) 25.
———. "Palabras a los intelectuales," 1961, *Ministerio de cultura, República de Cuba* Havana Cuba <http://www.min.cult.cu/historia/palabrasalosintelectuales.html>.
Corrales, José. "Los acosados, Tabo, Tota, Montes Huidobro y Piñera." *Círculo: Revista de cultura* 29 (2000):114–125.
Cortázar, Julio. *Rayuela.* Madrid: Cátedra, 2004.
Cortés, Jason. "La teatralización de la violencia y la complicidad del espectáculo de *Información para extranjeros* de Griselda Gambaro." *Latin American Theatre Review* 35.1 (2001): 47–61.

Cypess, Sandra, "Spanish American Theatre in the Twentieth Century." *The Cambridge History of Latin American Literature. Volume 2: The Twentieth Century.* ed. Roberto González Echevarría and Enrique Pupo-Walker. Cambridge: Cambridge University Press, 1996. 497–525.

Dauster, Frank N. *Historia del teatro hispanoamericano, siglos XIX y XX.* México: Ediciones de Andrea, 1973.

De Toro, Fernando. *Brecht en el teatro hispanoamericano.* Buenos Aires: Editorial Galerna, 1987.

Díaz Martínez, Manuel. "El caso Padilla: crimen y castigo (Recuerdos de un condenado)." *Inti: Revista de literatura hispánica* 46–47 (Otoño 1997-Primavera 1998): 157–166.

Diccionario de la lengua española. Ed. La Real Academia Española <http://www.rae.es/>.

Discépolo, Armando. *Stefano.* In *El grotesco criollo: Discépolo-Cossa.* Buenos Aires: Ediciones Colihue, n.d.

Dopico Black, Georgina. "The Limits of Expression: Intellectual Freedom in Postrevolutionary Cuba." *Cuban Studies/Estudios cubanos* 19 (1989), 107–142.

Dorfmann, Ariel. *La muerte y la doncella.* Buenos Aires: Ediciones de la Flor, 1992.

Driskell, Charles B. "Power, Myths and Aggression in Eduardo Pavlovsky's Theater." *Hispania: A Journal Devoted to the Teaching of Spanish and Portuguese* 65.4 (1982): 570–579.

Dubatti, Jorge. "Estudio preliminar." *Teatro completo I.* De Eduardo Pavlovsky. Buenos Aires: Atuel, 1997.

———. "Estudio preliminar." *Teatro completo II.* De Eduardo Pavlovsky. Buenos Aires: Atuel, 1998.

duBois, Page. *Torture and Truth.* New York:Routledge, 1991.

Echeverría, Esteban. "El matadero." *Antología del cuento hispanoamericano.* Ed. Fernando Burgos. México: Editorial Porrúa, 1991. 1–16.

Eidelberg, Nora. *Teatro experimental hispanoamericano, 1960–1980: La realidad social como manipulación.* Minnesota: Institute for the Study of Ideologies and Literature, 1985.

Esslin, Martin. *The Theatre of the Absurd.* Woodstock, NY: Overlook Press, 1973.

Escarpanter, José A. "Tres dramaturgos del inicio revolucionario: Abelardo Estorino, Antón Arrufat y José Triana." *Revista Iberoamericana* 56 (1990): 881–896.

Espinosa, Carlos. *Virgilio Piñera en persona.* Havana: Ediciones Unión, 2003.

Estorino, Abelardo. *La dolorosa historia del amor secreto de don José Jacinto Milanés. Teatro cubano contemporáneo.* Madrid: Fondo de Cultura Económica, 1992. 345–456.

———. *Memorias de Milanés.* Matanzas, Cuba: Ediciones Matanzas, 2005.

———. Personal interview. May 8, 2007.

Fanon, Frantz. *Wretched of the Earth.* Trans. Constance Farrington. New York: Grove Press, 1963.

Febles, Jorge. "Recontextualización poemática en *La dolorosa historia del amor secreto de don José Jacinto Milanés.*" *Latin American Theatre Review* 31.2 (1998), 79–95.
Feitlowitz, Marguertite. "Crisis, Terror, Disappearance: The Theater of Griselda Gambaro." In *Information for Foreigners.* Illiniois: Northwestern University Press, 1992.
Fernández, José M. "Teatro Experimental entrevista con Vicente Revuelta." *Conjunto* 3 (1964): 58–62.
Fleming, John. "Argentina on Stage: Griselda Gambaro's *Information for Foreigners* and *Antígona furiosa.*" *Selected Proceedings: Louisiana Conference on Hispanic Languages and Literatures* Ed. Joseph V. Ricapito. Baton Rouge: Louisiana State University Press, 1994.
Ford, Katherine. "El espectáculo revolucionario: El teatro cubano de la década de los sesenta." *Latin American Theatre Review* 39 (2005): 95–114.
Fornet, Ambrosio. "El Quinquenio gris: Revisitando el término." Havana 2006. http://www.criterios.es/pdf/fornetquinqueniogris.pdf
Forster, Merlin H. "Games and Endgames in Virgilio Piñera's *Dos viejos pánicos.*" *In Retrospect: Essays on Latin American Literature.* Ed. Elizabeth S. Rogers and Timothy J. Rogers. South Carolina: Special Literary Publications, 1987.
Foster, David William. *The Argentine Teatro independiente, 1930–1955.* South Carolina: Spanish Literature Publishing Company, 1986.
———. *Violence in Argentine Literature: Cultural Responses to Tyranny.* Columbia/London: University of Missouri Press, 1995.
Foucault, Michel. *Discipline & Punish: The Birth of the Prison.* Trans. Alan Sheridan. New York: Vintage-Random House, 1977.
Freire, Paolo. *Pedagogy of the Oppressed.* Trans. Myra Bergman Ramos. New York: Continuum, 2000.
Fuegi, John. *The Essential Brecht.* Los Angeles: Hennessey and Ingalls, 1972.
Gambaro, Griselda. *Information for Foreigners.* Ed. and Trans. Marguerite Feitlowitz. Illinois: Northwestern University Press, 1992.
———. *Teatro 2. Dar la vuelta. Información para extranjeros. Puesta en claro. Sucede lo que pasa.* Buenos Aires: Ediciones de la Flor, 1987.
García Chichester, Ana. "Virgilio Piñera and the Formulation of a National Literature." *CR: The New Centennial Review* 2.2 (2002): 231–251.
Garzón Céspedes, Francisco. "El teatro es un acto de solidaridad." *Conjunto* 60 (1984): 95–98.
Gerdes, Dick. "Recent Argentine Vanguard Theatre: Gambaro's *Información para extranjeros.*" *Latin American Theatre Review* 11.2 (1978): 11–16.
Gilman, Claudia. *Entre la pluma y el fusil: debates y dilemmas del escritor revolucionario en América Latina.* Siglo XXI Editores Argentina, 2003.
Girard, René. *Violence and the Sacred.* Trans. Patrick Gregory. Baltimore: Johns Hopkins University Press, 1977.
Giunta, Andrea. *Vanguardia, internacionalismo y política: Arte argentino en los años sesenta.* Buenos Aires: Editorial Paidós, 2001.
Goldman, Dara. "Los límites de la carne: Los cuerpos asediados de Virgilio Piñera." *Revista iberoamericana* 69 (2003): 1001–1015.
González, Aníbal. *Killer Books: Writing, Violence, and Ethics in Modern Spanish American Narrative.* Austin TX: University of Texas Press, 2001.

González Echevarría, Roberto. "Criticism and Literature in Revolutionary Cuba." *Cuba, Twenty-five Years of Revolution, 1959–1984.* Ed. Sandor Halebsky and John M. Kirk. New York: Praeger Publishers, 1985. 154–173.

Guevara, Ernesto Che. "El hombre nuevo." *Ideas en torno de Latinoamérica*, Vol I. México: Universidad Nacional Autónoma de México, 1986.

Halperín Donghi, Tulio. *The Contemporary History of Latin America.* Ed. and Trans. John Charles Chasteen. Durham: Duke University Press, 1993.

Henríquez Ureña, Max. *Panorama histórico de la literatura cubana.* La Habana: Editorial Arte y Literatura, 1978;

Historia de la Argentina. Buenos Aires: Crítica, 2002.

Historia de la literatura cubana: Tomo 1: La colonia: desde los orígenes hasta 1898. La Habana: Editorial Letras Cubanas, 2002.

Historia del Caribe. Barcelona: Editorial Crítica, 2001.

Hobsbawm, Eric. *The Age of Extremes: A History of the World, 1914–1991.* New York: Vintage-Random House, 1996.

Homenaje a Virgilio Piñera. Spec. issue of *Encuentro de la cultura cubana.* 20 (2001): 29–37.

Howell, Susana. "*Historias para ser contadas* y *Dos viejos pánicos*: Crítica y espectáculo." *Prismal/Cabral: Revista de literatura hispánica/Caderno afro-brasileiro asiático lusitano* 1 (1977): 17–24.

Huerta, Jorge. *Chicano Theater: Themes and Forms.* Tempe, Arizona: Bilingual Press/Editorial Bilingüe, 1982.

Kapcia, Antoni. *Havana: The Making of the Cuban Culture.* Oxford: Berg, 2005.

Kerrigan, Anthony. "What Are the Newly Literate Reading in Cuba? An 'Individualist' Memoir." *Salmagundi* 82–83 (1989): 322–335.

Jehenson, Myriam Yvonne. "Staging Cultural Violence: Griselda Gambaro and Argentina's 'Dirty War'." *Mosaic: A Journal for the Interdisciplinary Study of Literature* 32.1 (1999): 85–104.

Lane, Jill. *Blackface Cuba, 1840–1895.* Philadelphia: The University of Pennsylvania Press, 2005.

Larson, Catherine and Margarita Vargas, ed. *Latin American Women Dramatists: Theater, Texts, and Theories.* Bloomington, Indiana: Indiana University Press, 1998.

Leal, Rine. *Breve historia del teatro cubano.* La Habana: Editorial Letras Cubanas, 1980.

———. "Piñera todo teatral." Introducción. *Teatro completo.* De Virgilio Piñera. La Habana: Editorial Letras Cubanas, 2002. v–xxxiii.

Lindenberger, Herbert. *Historical Drama: The Relation of Literature and Reality.* Chicago: The University of Chicago Press, 1975.

Luzuriaga, Gerardo, ed. *Popular Theater for Social Change in Latin America: Essays in Spanish and English.* Los Angeles: UCLA Latin American Center Publications, 1978.

Lyday, Leon F. "De rebelión a morbosidad: Juegos interpersonales en tres dramas hispanoamericanos." *Actas del sexto congreso internacional de hispanistas.* Toronto: University of Toronto, 1980.

Lyday, Leon F. and George W. Woodyard, ed. *Dramatists in Revolt: The New Latin American Theater.* Austin: University of Texas Press, 1976.

Martin, Eleanor Jean. "*Dos viejos pánicos*: A Political Interpretation of the Cuban Theater of the Absurd." *Revista/Review iberoamericana* 9 (1979): 50–56.
Martínez Tabares, Vivian. "La dolorosa búsqueda de los recuerdos." *Teatro cubano contemporáneo*. Madrid: Sociedad Estatal Quinto Centenario, 1992. 347–354.
Matas, Julio. "Theater and Cinematography." *Revolutionary Change in Cuba*. Ed. Carmelo Mesa-Lago. Pittsburgh: University of Pittsburgh Press, 1971.
Mazziotti, Nora, comp. *Poder, deseo, y marginación: Aproximaciones a la obra de Griselda Gambaro*. Buenos Aires: Puntosur Editores, 1989.
Milanés, José Jacinto. *El conde Alarcos*. In *Teatro del siglo XIX*. La Habana: Editorial Letras Cubanas, 1986.
———. "El mendigo." *Poesía cubana de la colonia*. Arias, Salvador, ed. La Habana: Editorial Letras Cubanas, 2002. 77.
Milgram, Stanley. *Obedience to Authority: An Experimental View*. New York, HarperCollins Publishers, 1974.
Miranda, Julio E. "Sobre el nuevo teatro cubano." *Nueva literatura cubana*. Madrid: Taurus Ediciones, 1971.
Molinero, Rita, ed. *Virgilio Piñera: La memoria del cuerpo*. San Juan, P.R. : Editorial Plaza Mayor, 2002.
Montes-Huidobro, Matías. "El discurso teatral histórico-poético de Abelardo Estorino: entre el compromiso y la subversión." *Alba de América* 9.16–17 (1991), 245–57.
———. "Psicoanálisis fílmico-dramático de *La mueca*." *Monographic Review/ Revista monográfica* 7 (1991): 297–314.
———. *Persona, vida y máscara en el teatro cubano*. Miami: Ediciones Universales, 1973.
———. *El teatro cubano en el vórtice del compromiso 1959–1961*. Miami: Ediciones Universales, 2002.
Morson, Gary Saul and Caryl Emerson. *Mikhail Bakhtin: Creation of a Prosaics*. Stanford: Stanford University Press, 1990.
Mulvey, Laura. "Visual Pleasure and Narrative Cinema." *Visual and Other Pleasures*. Bloomington IN: Indiana University Press, 1989.
Mundani, Liliana. *Las mascaras de lo siniestro. Escena política y escena teatral en Argentina: El caso Gambaro*. Córdoba: Alción Editora, 2002.
Muñoz, Elías Miguel. "Teatro cubano de transición (1958–1964): Piñera y Estorino." *Latin American Theatre Review* 19.2 (1986): 39–44.
Muñoz, José Esteban. *Disidentifications: Queers of Color and the Performance of Politics*. Minneapolis: University of Minnesota Press, 1999.
Neglia, Erminio. *Aspectos del teatro moderno hispanoamericano*. Bogotá: Editorial Stella, 1975.
Ocasio, Rafael. "Gays and the Cuban Revolution: The Case of Reinaldo Arenas." *Latin American Perspectives* 29.2 (2002): 78–98.
Ordaz, Luis, et al. *Historia del teatro argentino*. Buenos Aires: Centro Editor de América Latina, 1982.
Padilla, Heberto. *Fuera del juego*. Miami: Ediciones Universal, 1998.
———. *Provocaciones*. Madrid: Ediciones La Gota de Agua, 1973.

———. *Sent off the Field: A Selection from the Poetry of Heberto Padilla*. Trans. J. M. Cohen. Great Britain: Andre Deutsch, 1972.

Palls, Terry L. "El teatro del absurdo en Cuba: El compromiso artístico frente al compromiso político." *Latin American Theatre Review* 11.2 (1978): 25–32.

Paquette, Robert L. *Sugar is Made with Blood: The Conspiracy of La Escalera and the Conflict between Empires over Slavery in Cuba*. Middletown, CT: Wesleyan University Press, 1988.

Parker, Andrew and Eve Kosofsky Sedgwick, ed. *Performativity and Performance*. New York: Routledge, 1995.

Pavis, Patrice. *Dictionary of the Theatre: Terms, Concepts and Analysis*. Trans. Christine Shantz. Toronto: University of Toronto Press, 1998.

———. *Languages of the Stage: Essays in the Semiology of the Theatre*. New York: Performing Arts Journal Publications, 1982.

Pavlovsky, Eduardo. "Algunos conceptos sobre el teatro de vanguardia." *Teatro del '60*. Buenos Aires: Ediciones Letra Buena, 1992. 113–120.

———. *La mueca y Cerca*. Buenos Aires: Ediciones Búsqueda, 1988.

Pellettieri, Osvaldo. *Una historia interrumpida: Teatro argentino moderno (1949– 1976)*. Buenos Aires: Editorial Galerna, 1997.

Pellettieri, Osvaldo, ed. *El teatro y su mundo: Estudios sobre teatro iberoamericano y argentino*. Buenos Aires: Editorial Galerna, 1997.

Pellettieri, Osvaldo, dir. *Historia del teatro argentino en Buenos Aires: La segunda modernidad (1949–1976)*. Vol. 4. Buenos Aires: Galerna, 2001.

———. *Historia del teatro argentino en Buenos Aires: El teatro actual (1976–1998)*. Vol. 5. Buenos Aires: Galerna, 2001.

Peters, Edward. *Torture*. New York: Basil Blackwell, 1985.

Pérez-Stable, Marifeli. *The Cuban Revolution*. New York: Oxford University Press, 1993.

Piñera, Vigilio. "La isla en peso." *La vida entera*. La Habana: UNEAC, 1968.

———. *Teatro completo*. Ed. Rine Leal. La Habana: Editorial Letras Cubanas, 2002.

Plamper, Jan. "Abolishing Ambiguity: Soviet Censorship Practices in the 1930s." *Russian Review* 60.4 (2001): 526–544.

"Primer congreso nacional de educación y cultura." *Gaceta de Cuba* 90–91 (March-April 1971): 2–13.

Postma, Rosalea. "Space and Spectator in the Theatre of Griselda Gambaro: *Informacion para extranjeros*." *Latin American Theatre Review* 14.1 (1980): 35–45.

Quintero Herencia, Juan Carlos. *Fulguración del espacio: Letras e imaginación institucional de la Revolución cubana (1960–1971)*. Rosario, Argentina: Beatriz Viterbo Editora, 2002.

de Quinto, José María. *El teatro iberoamericano en la década de los sesenta*. Murcia, Spain: Escuela Superior de Arte Dramático, 2000.

Quiroga, José. "Fleshing Out Virgilio Piñera from the Cuban Closet." *¿Entiendes? Queer Readings, Hispanic Writings*. Ed. Emilie L. Bergmann and Paul Julian Smith, Durham, N.C.: Duke University Press, 1995.

Quiroga, José. "Piñera inconcluso." *Virgilio Piñera: La memoria del cuerpo.* Molinero, Rita, ed. San Juan, P.R. : Editorial Plaza Mayor, 2002.

———. "Virgilio Piñera: On the Weight of the Insular Flesh." *Hispanisms and Homosexualities.* Ed. Silvia Molloy, Durham, N.C.: Duke University Press, 1998.

Reed , Roger. *The Cultural Revolution in Cuba.* Geneva: Latin American Round Table, 1991.

Rizk, Beatriz J. *El nuevo teatro latinoamericano: Una lectura histórica.* Minneapolis: Institute for the Study of Ideologies and Literature, 1987.

Rock, David. *Argentina 1516–1982: From Spanish Colonization to the Falklands War.* Berkeley: University of California Press, 1985.

Romero, José Luis. *Breve historia de la Argentina.* Buenos Aires: Fondo de Cultura Económica, 1996.

Rotker, Susana, ed. *Ciudadanías del miedo.* Caracas: Editorial Nueva Sociedad, 2000.

Ruiz del Vizo, Hortensia. "Análisis del miedo en *Dos viejos pánicos* de Virgilio Piñera." *National Symposium on Hispanic Theatre.* Ed. Adolfo M. Franco. Cedar Falls: University of Northern Iowa, 1985.

Sánchez-Grey Alba, Esther. "La obra de Virgilio Piñera, un hito en la dramaturgia cubana." *Círculo: Revista de cultura* 28 (1999): 50–60.

———. *Teatro Cubano Moderno: Dramaturgos.* Miami: Ediciones Universal, 2000.

Sandoval Sánchez, Alberto. "Bases teóricas para una lectura ex-céntrica del discurso colonial de Juan Ruiz de Alarcón." *Relecturas del Barroco de Indias.* Ed. Mabel Moraña. Hanover, NH: Ediciones del Norte, 1994. 281–302.

Santí, Enrico Mario. *Bienes del siglo: Sobre cultura cubana.* México: Fondo de Cultura Económica, 2002.

Scarry, Elaine. *The Body in Pain: The Making and Unmaking of the World.* New York: Oxford University Press, 1985.

Schanton, Pablo. "El teatro de la herida absurda." *Clarín: Revista de Cultura* 31 enero 2004, Ñ 6–10.

Schanzer, George O. "El teatro vanguardista de Eduardo Pavlovsky." *Latin American Theatre Review* 13.1 (1979) 5–13.

Schechner, Richard. *Performance Theory.* New York: Routledge, 2003.

Schutzman, Mady. "Calculus, Clairvoyance, and Communitas." *Women and Performance: A Journal of Feminist Theory* 10 (1999): 117–33.

Shakespeare, William. *Othello.* New Cork: Penguin Books, 1986.

Shoaf, Kristin E. "Los ciclos herméticos y el miedo al miedo en *Dos viejos pánicos* de Vigilio Piñera." *RLA: Romance Languages Annual* 9 (1997): 689–92.

Skinner, Eugene R. "Education and Theater in Post-Revolutionary Cuba." *Popular Theater for Social Change in Latin America: Essays in Spanish and English.* Ed. Gerardo Luzuriaga. Los Angeles: UCLA Latin American Center Publications, 1978.

———. "Research Guide to Post-Revolutionary Cuban Drama." *Latin American Theatre Review.* 7.2 (1974): 59–68.

Smith, Paul Christopher. "Theatre and Political Criteria in Cuba: Casa de las Américas Awards, 1960–1983." *Cuban Studies/Estudios cubanos* 14.1 (1984) 43–47.

Tarantuviez, Susana. *La escena del poder. El teatro de Griselda Gambaro*. Buenos Aires: Corregidor, 2007.

Taylor, Diana. *Disappearing Acts: Spectacles of Gender and Nationalism in Argentina's "Dirty War."* Durham: Duke University Press, 1997.

———. "Opening Remarks." *Negotiating Performance: Gender, Sexuality, and Theatricality in Latin/o America*. Taylor, Diana and Juan Villegas, ed. Durham: Duke University Press, 1994.

———. "Paradigmas de crisis: La obra dramatic de Griselda Gambaro." *En busca de una imagen: Ensayos críticos sobre Griselda Gambaro y José Triana*. ed. Diana Taylor. Canadá: Girol Books, 1989.

———. "Theater and Terrorism: Griselda Gambaro's 'Information for Foreigners'." *Theatre Journal* 42.2 (1990): 165–182.

———. *Theatre of Crisis: Drama and Politics in Latin America*. Kentucky: The University Press of Kentucky, 1991.

Taylor, Diana and Roselyn Costantino. *Holy Terrors: Latin American Women Perform*. Durham: Duke University Press, 2003.

Teatro argentino contemporáneo. Madrid: Sociedad Estatal Quinto Centenario, 1992.

Teatro cubano contemporáneo. Madrid: Sociedad Estatal Quinto Centenario, 1992.

Terán, Oscar. *Nuestros años sesenta: La formación de la nueva izquierda intelectual argentina, 1956–1966*. Buenos Aires: Ediciones El Cielo por Asalto, 1993.

Thomas, Hugh. *Cuba, or, The Pursuit of Freedom*. New York: Da Capo Press, 1998.

———. "Capítulo 2: La colonia española de Cuba." In *Historia del Caribe*. Barcelona: Editorial Crítica, 2001.

Toro, Alfonso de. "El teatro *menor* postmoderno de Eduardo Pavlovsky o el 'Borges/Bacon' del teatro: De la periferia al centro." Gestos 31 (Abril 2001): 99–110.

Torre, Juan Carlos and Liliana Riz. "Capítulo 7: Argentina desde 1946." *Historia de la Argentina*. John Lynch, et al. Barcelona: Crítica, 2001.

Torres Monreal, Francisco. "Introducción." *Teatro pánico*. By Fernando Arrabal, Madrid: Cátedra, 1986.

Triana, José. *La noche de los asesinos*. Madrid: Cátedra, 2001.

Usigli, Rodolfo. *El gesticulador/La mujer no hace milagros*. Mexico: Editores mexicanos unidos, 2002.

Versényi, Adam. *Theatre in Latin America: Religion, Politics, and Culture from Cortés to the 1980s*. Cambridge, Cambridge University Press, 1993.

Weiss, Judith A. *Casa de las Américas: an intellectual review in the Cuban Revolution*. Chapel Hill, North Carolina, 1977.

Woodyard, George. "The Theatre of the Absurd in Spanish America." *Comparative Drama* 111.3 (1969): 183–192.

———. "Estorino's Theatre: Customs and Conscience in Cuba." *Latin American Literary Review* 11.22 (1983): 57–63.

Zalacaín, Daniel. *El teatro absurdista hispanoamericano*. Valencia, España: Ediciones Albatros Hispanófila, 1985.

Index

Abdala, xi
Absurdist Theater, xiv–xv, 34
Aguilú de Murphy, Raquel, xiv, 34
Aire frío, 33
A la diestra de Dios Padre, xvi
Albuquerque, Severino João, 9, 43–44, 56
Alianza argentina anticomunista (Triple A), 102, 140
Alienation, *see* Distancing
Allende, Salvador, 11
Anderson, Thomas F., 32
Andreu, Olga, 63
Aramburu, General Pedro Eugenio, 100, 101, 140
Arenas, Reinaldo, 14, 30
Arendt, Hannah, 5–6
Areyto, xviii
Aristotle, 5
Arrabal, Fernando, 38–40, 44, 46, 48–49
Arrufat, Antón, xiv, 12, 13, 15, 16, 26, 27, 30, 31, 60, 61, 63, 145, 153, 163
Artaud, Antonin, xiii–xiv, xvii 3, 33, 43–44, 56, 176
Avelar, Idelber, 9, 148
Ávila, Leopoldo, 15–16, 61
Aztecs, ix

Bakhtin, Mikhail, 118, 120, 122
The Bald Soprano, xiv, 34, 36
Barranca abajo, xii
Batista, Fulgencio, 26

Beckett, Samuel, xiv, 34, 35
Belavel, Emilio, xviii
Berman, Sabina, xix
Biron, Rebecca, 9
Blanco, Roberto, 20, 57, 64
Boal, Augusto, xvi–xvii, 3, 4, 179
The Body in Pain, 6
Borges, Jorge Luis, 31
Brecht, Bertolt, xv–xvi, 3, 138, 163, 172
Brene, José, 27
Breton, Andre, 39
Bruguera, Tania, xix
Buenaventura, Enrique, xvi
Butler, Judith, 125, 129–130

Cabildo Teatral, 27
Cabrera Infante, Guillermo, 12, 15, 29, 40, 61
Calvo, César, 12
Campbell, Joseph, 39
Cámpora, Héctor, 102, 140
Camus, Albert, xiii, 33
Carballido, Emilio, xiii, xv
Carpentier, Alejo, 26
Carrió Mendía, Raquel, 33
Casa de Comedias, xi see also *Teatro de la Ranchería*
Casa de las Américas, 30, 33, 60, 61, 100
Casal, Lourdes, 15
Casas, Myrna, xviii
Caso Padilla, 2–3, 11–18, 28, 32, 57, 58, 60, 177 *see also* Padilla Affair

Castellanos, Rosario, xviii
Castro, Fidel, 1, 11, 13, 14, 16, 18, 28, 29, 30, 40
Castro, Raúl, 1, 60, 61
Censorship in Cuba, 28–30, 31, 54, 66
Censorship, Soviet, 29
Chávez, César, xix
Chingana, xi
Ciclón, 32
Cohen, J.M., 12
Condorcanqui, x
Congress on Education and Culture, 58, 60
Conjunto Dramático de Oriente, 27
Conspiración de la Escalera, 58, 59, 65, 67, 84–94
Copernicus, 17
Cordobazo, 101, 103, 124
Corrales, José, 37, 38
Cortázar, Julio, 153
Cortés, Hernán, ix
Cortés, Jason, 153, 158
Cossa, Roberto, 100
Costantino, Roselyn, xviii
Cristal roto en el tiempo, xviii
Cuban Communist Party, 26
Cuban Missile Crisis, 29
Cuban Revolution, xvi, 11, 14, 15, 16, 17, 18, 20, 26, 27, 32, 34–35, 44, 48, 58, 59, 60, 63, 66, 73, 79, 94
Cuentos fríos, 31
Cuza Malé, Belkis, 13
Cuzzani, Agustín, xv

Dadaists, 39
de Beauvoir, Simone, 13
de la Cruz, Sor Juana Inés, x, xviii
del Monte, Domingo, 62, 77, 81–82
Del sol naciente, 139
de Palma, Ramón, 81–82

de Toro, Alfonso, 110, 120, 122
de Toro, Fernando, xv
de Zequeira y Arango, Manuel, 74–76
Díaz, Jorge, xiv, 35
Díaz Martínez, Manuel, 12
Díaz Ordaz, Gustavo, 11, 26
Dictionary of the Real Academia Española, 38 105, 106, 142
Dirty War, *see* Proceso de Reorganización Nacional
Discépolo, Armando, xii
Distancing, xv, 138
Dos viejos pánicos, 8, 19–20, 25–56, 58, 97, 176, 178
Dragún, Osvaldo, xv
Driskell, Charles B., 105
Dubatti, Jorge, 99, 113
duBois, Page, 90

Eidelberg, Nora, xvi
Ejército Revolucionario del Pueblo (ERP), 102, 124, 140
El caimán bardudo, 12
El campo, 139
El cepillo de dientes, xiv
El conde Alarcos, xi, 63, 64, 82–83
Electra Garrigó, 32, 36
El eterno femenino, xviii
El gesticulador, xii–xiii
"El mendigo," 71–73
El Mendigo, 64–65, 69, 70–78, 80, 137
El no, 33
El robo del cochino, 27, 61
El señor Galíndez, 99
"El socialismo y el hombre en Cuba," 4
Eltit, Diamela, xix
Encuentro de la cultura cubana, 31
Epic Theater, xv–xvi
Episches Theater, see Epic Theater
Espuela de Plata, 31
Esslin, Martin, xiii, 33, 110

INDEX

Estorino, Abelardo, 19, 20, 27, 57–95, 137, 144, 145, 175, 176, 177, 178

Falkland Islands War, see La Guerra de las Islas Malvinas
Falsa alarma, 36
Fanon, Frantz, 5, 6
Febles, Jorge, 63
Ferrigno, Oscar, 100
Fornet, Ambrosio, 60–61
Forster, Merlin H., 37, 38, 45
Foucault, Michel, 4, 5, 148, 163–164, 177
Freire, Paulo, 2
Frondizi, Arturo, 100–101
Fuerzas Armadas Revolucionarias, 61

Galileo, 16–17
Gambaro, Griselda, xiv, xviii, 19, 21–22, 137–173, 175, 176, 177, 178
García Chichester, Ana, 32
García Márquez, Gabriel, 13
Gerdes, Dick, 147
Gilman, Claudia, 123–124
Girard, René, 5, 55, 84, 176–177
Gombrowicz, Witold, 31
González, Aníbal, 9
Gorostiza, Celestino, xiii
Grotesco criollo, xii
Grupo de Actores Profesionales (G.A.P.), 100
Grupo de estudios del teatro argentino (GETEA), 141
Guevara, Che, 1, 4, 11, 26
Guido, José María, 101
Guillén, Nicolás, 26
Gutiérrez, Eduardo, xii

Heredia, José María, 81–82
Herlinghaus, Hermann, 9–10
Hernández Espinosa, Eugenio, 27
Hobsbawm, Eric, 10–11

Hubert de Blanck, 64
Huerta, Jorge, xix

Illia, Arturo, 101
Indigenous theater, ix–x
Información para extranjeros, 8, 21–22, 137–173, 176, 178
Institute of the Book, 29
Ionesco, Eugene, xiv, 34, 35

Jodorowsky, Alejandro, 38
Juan Moreira, xii
Juventud Peronista, 102

Kriegar Vasena, Adalbert, 101

La Bamba, xix
La caja de zapatos vacía, 33, 46
La carne de René, 31
La dolorosa historia del amor secreto de don José Jacinto Milanés, 20, 57–95, 137, 144, 145, 176, 178
La Gaceta de Cuba, 16, 17
La Guerra de las Islas Malvinas, 139
La isla en peso, 31, 32
La malasangre, 139
La mueca, 20–21, 97–135, 176, 178
Lane, Jill, xi, 87
La niñita querida, 33
La noche de los asesinos, 84
Lanusse, General Alejandro, 101–102, 139
Leal, Rine, 19, 28, 33
Lector cómplice, 153
Levingston, General Roberto Marcelo, 101–102, 140
Lezama Lima, José, 12, 30, 32, 66
Lindenberger, Herbert, 94
Lockwood, Lee, 15
Lope de Vega, 63
López, César, 16
Los mangos de Caín, 61
Los siameses, 139
Lunes de Revolución, 40

218   INDEX

Manzano, Juan Francisco, 87
Marqués, René, xv
Martí, José, xi
Martin, Eleanor Jean, xiv, 34–35
Martínez de Perón, María Estela "Isabel," 102, 103, 140–141
Martínez Tabares, Vivian, 63
Milanés, Carlota, 70–71, 78, 79–80
Milanés, Federico, 65, 78–80
Milanés, José Jacinto, xi, 20, 57, 58, 59, 62–87, 92–94, 137
Milgram, Stanley, 154
Mira de Mescua, 63
Molinero, Rita, 31
Montes Huidobro, Matías, 37, 67–68, 70, 73, 77–78, 98–108
Montoneros, 101, 102, 124, 140
Mulvey, Laura, 108–110

National Council of Culture, 61
Nuevo Teatro Popular, xvi

Ocampo, Victoria, 31
Ollantay, x–xi
Onganía, General, 101
On Violence, 5
Orígenes, 31
Otero, Lisandro, 12
Othello, 166–168, 170, 171

Padilla Affair, 2–3, 11–18, 28, 32, 57, 58, 60, 177 see also caso Padilla
Padilla, Heberto, 12–18, 26, 27, 30, 60, 61, 66, 91, 92, 94, 177
Palls, Terry, xiv, 34
Pan, 38–40
Panic Theater, 38–40, 44, 46, 47, 48–49
Pavis, Patrice, 21, 145
Pavlovsky, Eduardo, 12, 20–21, 97–135, 139, 175, 176, 177, 178
Pavón Tamayo, Luis, 61
Pavonato, 61
Paz, Octavio, 13, 28, 60

Pedagogia do oprimido, 2
Performance, 21–22, 138
Perón, Isabel, see Martínez de Perón, María Estela "Isabel"
Peronism, 100–103, 140
Perón, Juan Domingo, 100–103, 140
Piñera, Virgilio, xiv, 1, 19–20, 25–56, 61, 63, 64, 97, 175, 176, 177, 178
Plácido, 59, 65, 84–87, 91–92
Planilla, 25, 26, 47–51, 52
Poesía en Voz Alta, xviii
Ponte, Antonio José, 31
Postma, Rosalea, 145, 148, 150, 171
Proceso de Reorganización Nacional, 99, 103 141, 173

Quinquenio gris, 14, 32, 60–61

Raznovich, Diana, xix
Reed, Roger, 16–17
Revolución, 40
Revolution of 1959, see Cuban Revolution
Revuelta, Vicente, 4, 64
Rock, David, 102
Rodríguez, Jesusa, xix
Rotker, Susana, 9
Rozenmacher, Germán, 100
Ruiz de Alarcón, Juan, x

Sánchez, Florencio, xii
Santí, Enrico Mario, 32
Sartre, Jean Paul, xiii, 13, 28, 33, 60
Scarry, Elaine, 6, 156, 177
Schanzer, George O., 107
Schechner, Richard, 106 see also caso Padilla
Solórzano, Carlos, xv
Somigliana, Carlos, 100
Somos, 98
Spectacularity, 2, 18
Stalinism, 14, 17, 29, 124
Stefano, xii
Storni, Alfonsina, xviii

INDEX

Sur, 31
Surrealists, 39

Tacón Theater, 64
Tallet, José Z., 12
Tarantuviez, Susana, 138, 145
Taylor, Diana, ix, xviii–xix, 21–22, 35
Teatro Abierto, 100
Teatro bufo, xi, 35
Teatro Campesino, xix
Teatro de la Ranchería, xi *see also* Casa de Comedias
Teatro del oprimido, 2, 3
Teatro Escambray, 27
Teatro Estudio, 4, 61
Teatro Experimental de Cali (TEC), xvi
Teatro gaucho, xii
Teatro Irrumpe, 20, 57, 64
Teatro Orientación, xiii
Theater of Cruelty, xiii, xvii, 33, 43, 56, 145
Theater of the Absurd, xiii–xv, 19, 33–36, 37, 47, 98
Theatre of crisis, 35
Theatre of the Oppressed, xvi–xvii
Thomas, Hugh, 28
Topor, Roland, 38
Torres Monreal, Francisco, 48, 49

Triana, José, 26, 44, 46, 63, 84
Tzu, Sun, 148

Ulises, xiii
Un hogar sólido, xviii
Unión Nacional de Escritores y Artistas de Cuba (UNEAC), 12, 13, 16, 17, 28, 30
Usigli, Rodolfo, xii–xiii

Verde olivo, 15, 16, 61
Verfumdung, see Distancing
Versényi, Adam, ix–xii
Viborazo, 140
"Vida de Flora," 31, 32
Villaurrutia, Xavier, xiii
Villaverde, Cirilo, 81–82
von Clausewitz, Carl, 148

Waiting for Godot, xiv, 34
Weiss, Judith A, xvi
Wolff, Egon, 100
Woodyard, George, xiv, xv, 34, 47

Ximeno, Isabel, 62, 65

Zafra de los diez millones, 29, 59
Zalacaín, Daniel, xiv, 34
Zoot Suit, xix